THE URBAN NATURALIST

THE URBAN NATURALIST

HOW TO MAKE THE CITY YOUR SCIENTIFIC PLAYGROUND

MENNO SCHILTHUIZEN

ILLUSTRATIONS BY JONO NUSSBAUM

THE MIT PRESS   CAMBRIDGE, MASSACHUSETTS   LONDON, ENGLAND

The MIT Press
Massachusetts Institute of Technology
77 Massachusetts Avenue, Cambridge, MA 02139
mitpress.mit.edu

The MIT Press would like to thank the anonymous peer reviewers who provided comments on drafts of this book. The generous work of academic experts is essential for establishing the authority and quality of our publications. We acknowledge with gratitude the contributions of these otherwise uncredited readers.

This book was set in ITC Stone and Avenir by New Best-set Typesetters Ltd. Printed and bound in the United States of America.

Library of Congress Cataloging-in-Publication Data is available.

ISBN: 978-0-262-04909-2

10 9 8 7 6 5 4 3 2 1

EU product safety and compliance information contact is: mitp-eu-gpsr@mit.edu

publication supported by a grant from
The Community Foundation for Greater New Haven
as part of the *Urban Haven Project*

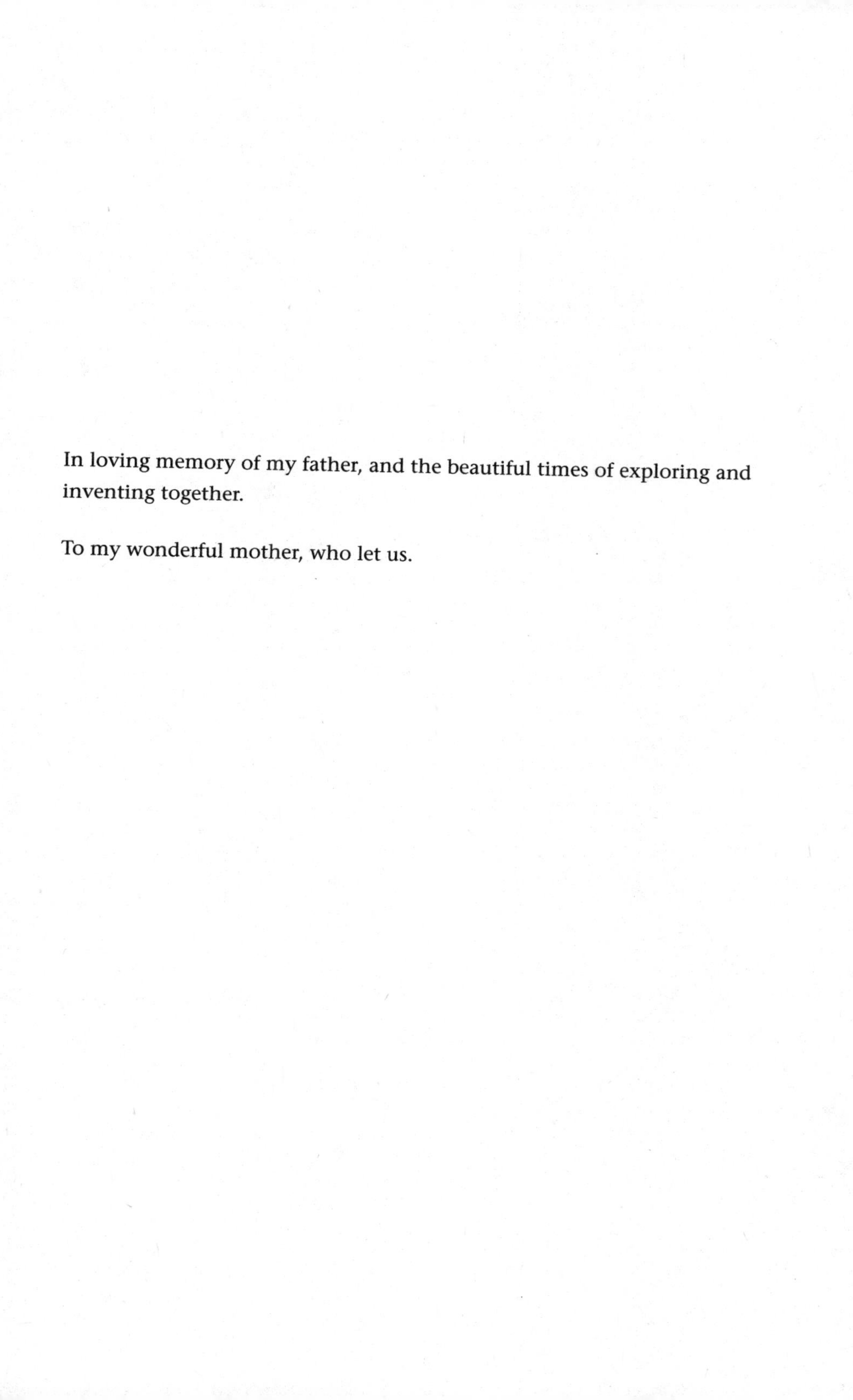

In loving memory of my father, and the beautiful times of exploring and inventing together.

To my wonderful mother, who let us.

CONTENTS

PREFACE

LOSING GROUND

My dad's expensive Maxim piezo cigarette lighter, a work anniversary gift from his colleagues, hurtled across our living room and hit the opposite wall with an ominous crack, sending a shower of shards from its black plastic casing flying off in various directions. Its battered remains came to rest on the carpet in front of the bookshelf. I looked at it in horror and immediately regretted having made it bear the brunt of my outrage.

Half an hour earlier, I had visited my favorite place: a small untended patch of earth close to our house. A collection of wildflowers grew there, and the whole summer I would wade through chest-high cow parsley, purple loosestrife, and white dead-nettle. I called it my "herb garden." It was perhaps only fifty square meters, but to me it was my private jungle and the only piece of wilderness in an otherwise manicured suburban neighborhood near Rotterdam. I found my first twenty-two-spot lady-bird beetle here, and my first peacock butterflies—common insects, but to me, an eleven-year-old budding nature enthusiast, it was like finding elephants and lions. I also went there to cut grass for my three pet rabbits, which lived in a cage in our backyard.

But that day, the first thing I saw was a municipal maintenance vehicle and two men in coveralls who were busy spraying a white mist on my herb garden. Herbicide, I knew. To kill the rampant wildflowers. To make the men feel guilty, I demonstratively entered my herb garden and began cutting grass for my rabbits. "Hey, better stay away, buddy," one of them called at me as they were getting ready to drive away. "We've just sprayed that place with poison."

Enraged and powerless, I ran back to the house, grabbed the first object in view, and threw it with all my force against the wall. That evening, somewhat calmed down, I had to explain what happened and show my dad his mangled lighter. He made me buy him a new one from my pocket money and told me to learn to control my temper.

And although initially I did engage in a spree of juvenile environmental activism that led to a run-in with authorities far less benign than my father, and which I will divulge later in this book, I did eventually learn to channel my passion in a more mature way.

THE POWER OF COMMUNITY SCIENCE

Cut to forty-five years later: Leiden, a city between Amsterdam and The Hague. Different hometown, different urban nature, still the same dissatisfaction with municipal disregard for wild vegetation but with less destructive and more effective ways to deal with it. We have built a makeshift lab at Rudolf's houseboat. Rudolf Tenzer is a schoolteacher who, since the 1980s, has been nurturing a deserted strip between the railroad and the canal into a true urban wild forest—a tangle of brambles, elders, and alders. Three hundred meters long and only fifteen wide, devoid of paths or other infrastructure and quite inaccessible, it is too big to be a private garden, too awkwardly shaped and unofficial to be a city park, and therefore an easy target for the municipality, which has planned to pave it over.

Rather than brandishing banners and placards (let alone flinging vintage cigarette lighters), we have decided to use an altogether different weapon to fight for its survival: science. Rudolf has recruited a group of concerned neighborhood people and I have assembled a team of top-notch biodiversity experts. Each month this motley crew meets at Rudolf's houseboat to jointly study the animals, plants, and fungi that make up the ecosystem of this strip of urban forest. We use clip-on lenses to turn smartphones into microscopes and new image analysis apps that allow us to identify species from photographs.

Today we have invited the city council members to join us, and a few of them actually show up. We take them to the edge of the railroad track to introduce them to *Zodarium rubidum*, a small spider that lives among the pebbles. It has not been found in the Netherlands before. It does not build a web but hunts ants and carries their corpses around. This makes it smell like ant and thus evade detection by the other ants, so that it can eat . . . more ants. We pass around hand lenses for everyone to have a look. The council members gasp in amazement. Next we drag them to the ivy growing on the maple trees and bring out a device that resembles something from *Ghostbusters* (but is actually a heavy-duty petrol engine–powered vacuum cleaner to suck small insects from dense vegetation). After wielding the machine for a couple of minutes (the council members give it a try too), we sprinkle the collected material on a sheet, and among the insects and spiders that hurriedly scatter in all directions are a couple of tiny dark brown cockroaches with a delicate white line across the middle of their body. They're juveniles of *Planuncus tingitanus*, a special cockroach found only in cities. Next the local politicians are shown the empty snail shells in which *Osmia spinulosa*, the spined mason bee, another rarity, builds its nest.

Still reeling from the Latin names, the council members stagger back to their town hall. A few months later, the city announces that the plans for the new infrastructure, a glitzy bicycle path, have been altered so that it can run along an existing road, to protect the "special biodiversity" of Rudolf's urban jungle.

During my entire career of thirty-odd years, working as an academic in evolutionary biology, ecology, and biodiversity in a string of universities and research institutes, I had done dozens of research projects ranging from the evolution of land snails on Crete, to the penis shape of cave beetles, to the DNA of mountain birds in Borneo. I had always included a perfunctory paragraph in my research papers that the work would eventually go toward helping conserve the subject of my studies. But never, ever had I actually managed to conserve anything—until that day.

Working with laypeople eager to learn how to explore and understand their own environment, I had formed a team of urban naturalists who gathered scientific data that added knowledge and value to that environment, and joyously advertised their discoveries. The ant-hunting spider, the banded cockroach, and the shell-nesting mason bee became mascots of that narrow strip of urban forest and were highlighted on the website that the neighborhood had set up for its biodiversity project. Those rare species, but even more so the pride with which the neighborhood naturalists introduced them to the visiting council members, meant that the area graduated from disposable wasteland to cherished biodiversity hotspot, and was pardoned on the scaffold.

But more special than those invertebrate rarities were Rudolf and his band of honorary neighborhood scientists. It is tempting to refer to them as citizen scientists, but that term might imply that they were just there as assistants to me and the other "real" scientists. In fact, their reach stretched much further than that: they were the conceivers, initiators, executers, and reporters of their own research project. Their fieldsite was smack in the quarter of town where they had their homes and families. They would meet their friends on the street and talk about their latest discoveries, which would then be posted to the neighborhood WhatsApp group. In other words, they were *community scientists*. It was their own research project, and we professionals were only there for guidance.

This story shows that science has come full circle. It started out as an intellectual pastime for well-to-do ladies and gentlemen. In the nineteenth century, famous scientists such as Charles Darwin and the men and women in his network were priests, notaries, doctors, housewives, pensioners, schoolteachers, and lawyers. None was a professional scientist. They did their groundbreaking research at home and in their nurseries and gardens. Fieldwork was done during family outings, and papers were written by gaslight at night and read to a like-minded audience on evening gatherings of learned societies.

Then came the twentieth century, which saw a professionalization of science. As knowledge and technologies progressed, it was no longer easy for laypeople to make a meaningful contribution. You needed large labs full of students, technicians, and expensive equipment, funds amounting to millions, and huge libraries and computers to access and organize your data.

But since the 2000s, science has quietly regained access to those garden sheds and living rooms. Thanks to the open science revolution, the world's entire scientific literature is available online via Google Scholar and Sci-Hub. You can buy cheap PCR machines, Geiger counters, and even electron microscopes on eBay and Amazon. Huge amounts of scientific data, from genome sequences to soil pollution measurements, can be downloaded by all and processed by anyone at home on a personal computer with open-source software. The skills required to do all this can be learned on massive open online courses (MOOCs), and online forums, tutorials, and Facebook groups can help the beginner get started and the advanced community scientist get unstuck. Just as two centuries ago, high science is again accessible to anyone with a brain and a willingness to use it and to connect with the brainpower and hearts of others. You can be a nineteenth-century naturalist today.

More important, community science has revived the romance that professional science is risking to lose. Mired in top-down management, key performance indicators, and cutthroat competition for grants and publication in high-impact journals, many modern professional scientists have become academic managers, forced to leave the fun fieldwork to students, the lab experiments to research technicians, and the writing of articles to those nomadic short-term researchers called postdocs. Community scientists, on the other hand, can give free rein to their curiosity, enthusiasm, and, yes, the romance of doing science.

HOW TO BE AN URBAN NATURALIST

In the first part of this book, I introduce you to the tools of the trade of the urban community scientist. Though I will describe internet resources, low-tech gadgets, and off-the-shelf gizmos to do high-quality biology at home, my story is less about tricks of the trade and more about a new view of life that emerges once you embark on the path of the urban naturalist. I show the benefits of building a natural history collection, and discuss the ethics of taking samples from nature for knowledge. I also show how careful observation can help shape your ideas, generate new questions, and produce truly original science. I introduce a new band of people who have rekindled their interest in their immediate surroundings and have discovered the bliss of exploring what is nearby.

The second part, a series of loosely connected vignettes, is meant to promote the urban environment as the ideal field setting for the community naturalist. The unusual anthropogenic ecology of the city, its ecosystem assembled from native species mixed with escapees from our agricultural, pet, gardening, and aquarium trades, and its patchy geography crisscrossed by barriers and corridors both above- and belowground, offers a gold mine of settings for the curious naturalist. I will take you on adventures across the city and show how, to the modern naturalist, the urban habitat is every bit as exciting and unexpected as "wild" nature. As I discussed in my book *Darwin Comes to Town*, we are witnessing the evolution of an entirely new global ecosystem, created by the actions of a single species—namely, humans—a unique event in the history of life on Earth. In this new book, I provide examples from all over the world of how the urban naturalist can and does use the city to make fundamental and unexpected discoveries in biodiversity, behavior, ecology, and evolution.

In the third and last part of the book, I point out that cities are not just interesting as novel ecosystems. They are also where most people live, where impromptu communities can form around a shared love and concern for a local patch of urban nature. Through a multitude of causes, daily exposure to nature improves people's mental and physical health, but it is often forgotten that such exposure takes place in the mind as much as in the outdoors. By studying the natural history of the animals, plants, and fungi that often remain unseen to the untrained eye, urban naturalists become much more aware of the nature that exists all around us and can become storytellers in their community, making other people appreciate their surroundings better. In a world where cities are rapidly growing and more and more of us rarely visit "wild" nature anymore, being able to enjoy and protect urban nature will be essential to our well-being.

I also weave a few personal stories into the narrative. Though I was the first person in my family to have gone to university, I come from a lower-middle-class background where doing citizen science was a thing long before it was called that. My grandfather and father both held day jobs but filled their spare time with a broad repertoire of science-leaning hobbies, doing physics and chemistry experiments in the kitchen, collecting minerals and fossils, assembling electronic components to build bat detectors and highly experimental musical instruments, and creating telescopes and microscopes from old lenses. Thanks to them, I became imbued with a sense of wonder and adventure for anything to do with science. At a young age, I began an eclectic natural history collection, which eventually led to my professional career as a biologist. At some point during that career, that original passion became curtailed by the daily business of academic life. In this book, I describe how I created my own brand of science communication and community science and rekindled a kind of intellectual romance that I had very nearly lost. And I describe how, by working alongside community scientists and their infectious enthusiasm, I am falling in love with the naturalist's life all over again.

1

BUILD YOUR OWN DOWN HOUSE: WE CAN ALL BE VICTORIAN NATURALISTS NOW

I often think the more I limit myself to a small area, the more novelties and discoveries I make in natural history.
—Mary Treat, *Home Studies in Nature* (1885)

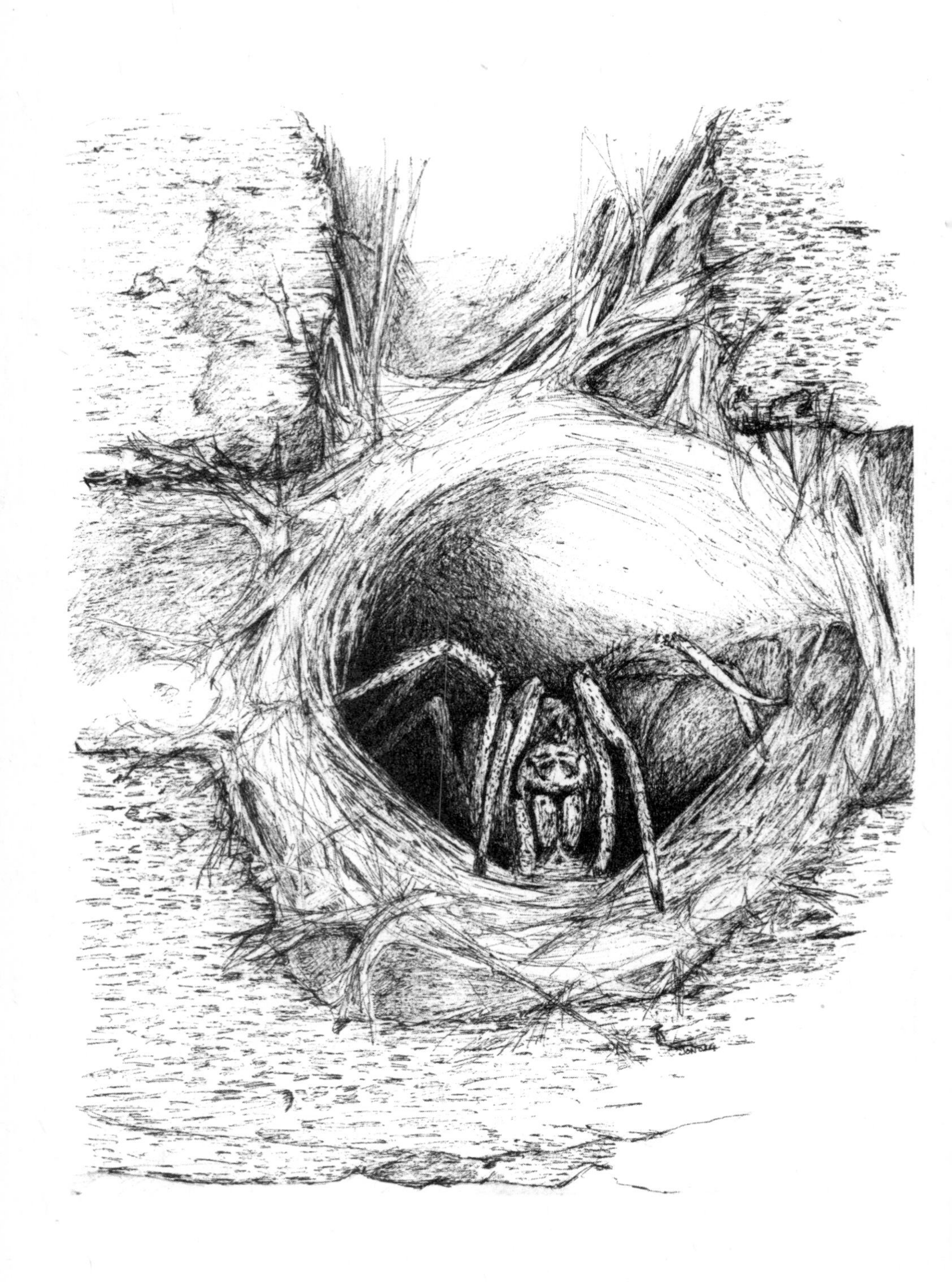

1

THE AGE OF THE AMATEUR

It is a curious and somewhat daring thing, that pen drawing that graces page 112 of the 1885 book *Home Studies in Nature*. We see a nearly closed circle of dense shrubs with a lawn in the center, in the middle of which stands a lone maple. In the shade cast by this tree is a seated figure, a lady in a long dark dress, wearing a white summer hat and light-colored shoes. The few pen strokes that hold her up give us little more information about who she is or what she is doing, except that she appears to be reading or writing. Peering even closer at the picture, behind her we discern a small foldable stool and even tinier objects, barely a millimeter tall in the printed image, scattered on the grass. They look like upturned bowls or cups. The title beneath the image reads, somewhat unexpectedly, "The Insect Menagerie."

That title, as the text on that page reveals, shows that in those days the term "insect" was applied more broadly because those upside-down bowls are glass bell jars that house not true six-legged insects but giant eight-legged wolf spiders. And the seated woman is the person who put them there for her home-bound scientific study and who, in the drawing, is probably busy writing field notes on her experiments. Her name: Mary Lua Adelia Treat.

Mary Treat was born in 1830, the daughter of an Ohio clergyman. With her husband, a doctor, she moved to Vineland, New Jersey, at the time a utopian settlement based on temperance and agriculture. Together, they dabble in horticulture and the study of insect pests, but in 1870, her husband leaves her. Rather than living up to the contemporary expectation of her settling in the quiet life of a middle-aged divorcee, Mary Treat blossoms, and launches into a wide range of scientific pursuits.

First, continuing the pest control interests of her married life, she publishes occasional articles on agricultural entomology, but soon her scientific exploits lead her into such diverse fields as butterfly metamorphosis, how birds learn to improve their nest-building skills, and the behavior of harvester ants. Her output also intensifies: between 1875 and 1890 she publishes nearly one hundred books and scientific articles—a publication record that many a modern-day biologist would be envious of.

As her bibliography grows, she develops her own style of writing, characterized by careful observations of the behaviors of animals *and* plants (she studies how carnivorous plants catch their prey), cutting-edge Darwinism (Darwin was one of her many correspondents), and also a personal, light-hearted bonding with her readers, with whom she shares her scientific adventures, most of which take place in her own home and garden. She publishes so as not to perish, metaphorically but also literally: as a single woman of slight financial means, she uses publishers' payments to keep her private science enterprise afloat. That is also why she chooses to send her work to popular literary magazines with large audiences, such as *Harper's Magazine*, rather than to scientific journals. In one letter to Darwin, she confides, "I am wholly dependent upon my own exertions, and must go where they pay best."

Over the years, Treat steadily gains prestige in the world of science. Living in a small town with no easy access to the learned societies of Philadelphia, New York, or Boston, she corresponds with a growing network of naturalists around the world, who help her with insights, information, and literature. The letters she exchanges in the 1870s with the elderly Darwin reveal an increasingly warm friendship in the course of which the two discuss not only natural history but also the growth of Darwinism in the United States, the frightfully hot weather in New Jersey, and Darwin's little ailments. Besides letters, the envelopes that cross the Atlantic also contain small gifts: herbarium specimens and Mary's and Charles's recent publications. Darwin receives reprints of Treat's articles and he in return mails her a complimentary copy of his latest book, *Insectivorous Plants*, to which Treat replies, "I was so fascinated with it that I sat up nearly all night before I could lay it down."

Others, like Missouri state entomologist Charles Valentine Riley, thirteen years her junior, feel the need to adopt a more condescending tone.

In reply to a letter in which Treat proposes a novel theory for how the sex of caterpillars may be determined, Riley replies, "More error and confusion creeps into our science by these rash and unequivocal conclusions than in any other way. . . . I am too busy now to write more fully; but I fear you will find, in the future, that you have spoken too implicitly on insufficient grounds; and you have no idea how much harm it may do." Granted, it later turned out that Treat was indeed wrong, but that is no reason to discourage a passionate naturalist so harshly.

Undaunted, Treat continues to build her own brand of natural history. Her home and garden experiments become more sophisticated, more daring, and she displays a particular predilection for the macabre, which undoubtedly gains her many readers. Carnivorous plants in particular take her fancy as they are "a class of plants that seem to have reversed the regular order of nature, and, like avengers of their kingdom, have turned upon animals, incarcerating and finally killing them." On one occasion, she tests the digestive powers of the Venus fly trap by letting it bite into the tip of her little finger. She intends to keep her finger inside the plant's jaws for five hours to see if any skin dissolution would take place, but "was surprised at the amount of pressure about my finger. . . . In less than two hours I was obliged to take my finger from the plant, defeated in so simple an experiment, and heartily ashamed that I could not better control my nerves."

One of her pet organisms is the carnivorous aquatic bladderwort, *Utricularia*. Bladderworts' underwater stems have tiny traps, which, as Treat discovered under her microscope, often contain the remains of crustaceans and larvae of aquatic insects. At the time, it was generally believed that the bladders or utricles were merely floating devices and that insects blundered into them and accidentally got caught inside. Darwin, who was simultaneously researching them for his book, once painstakingly removed all the bladders from a plant and saw that its buoyancy was unaffected, thus rejecting the life preserver hypothesis, but he also couldn't work out whether *Utricularia* bladders were actually catching prey, and, if so, how. In the end, he concluded that "the animals enter merely by forcing their way through the slit-like orifice, their heads serving as a wedge." But Treat, who kept a bladderwort in a tub of water and spent days "feeding" it small invertebrate animals such as water bears, "soon had the satisfaction of seeing the *modus operandi* by which the victim was caught."

As it turned out, a small depression in the bladder, the vestibule, has a valve that acts as a "fatal trap." She writes, "I was very much amused in watching a water-bear (*Tardigrada*) entrapped." 'It slowly walked around the utricle, as if reconnoitring—very much like its larger namesake; finally it ventured into the vestibule, and soon heedlessly touched the trap, when it was taken within so quickly that my eyes could not follow the motion. The utricle was transparent and quite empty, so that I could see the behavior of the little animal very distinctly. It seemed to look around as if surprised to find itself in so elegant a chamber; but it was soon quiet, and on the morning following it was entirely motionless, with its little feet and claws standing stiff and rigid. The wicked plant had killed it."

Darwin, when Treat writes to him about her results, graciously admits academic defeat: "It certainly appears from your excellent observations that the valve was sensitive. . . . It is pretty clear I am quite wrong. . . . The indraught of the [prey] is astonishing." Treat, in her book *Home Studies in Nature*, hypothesizes that the plant creates a partial vacuum inside the blister, and that the release of the valve creates enough flow to suck the insect in. "But how could a vacuum be formed?" Research in the 1940s provided the answer: the proteins in the cell walls of the blister constantly pump out water, creating a negative pressure inside.

The circular spider menagerie in her garden with which we began this chapter is another one of Treat's home-grown biology projects. Though mostly out of necessity, Treat emphasizes the benefits of doing research in one's immediate surroundings rather than in faraway places. "I often think the more I limit myself to a small area, the more novelties and discoveries I make in natural history," she writes in a *Harper's Magazine* article in which she tells us what she learned from her "insect menagerie."

To begin with, under each of those bell jars is one of her experimental wolf spider females, transported there by her to have them close by for observations. The enclosure keeps out any predators, such as large spider-hunting wasps. Through the glass, presumably on her knees on the lawn, Treat watches how the spiders construct their nests. These spiders (a new species that Treat discovered and named *Tarantula turricola*) do not build any webs but burrow into the ground and embellish the entrance of their burrow with a tidy turret of twigs, held together by silk.

Treat creates an audience for her scientific discoveries thanks to the chatty, evocative, humorous, but nonetheless very exact manner in which she describes her observations, and she goes all-out in the picture she paints of spider mothers' nest-building behavior. "She first spins a canopy of web over the funnel, leaving a place of exit on one side. She next comes out and steps carefully over the canopy, as if to see whether it is strong and secure. Seemingly satisfied that it is all right, she steps down, just letting her fore-feet touch the edge of the web, while with her hind-legs she feels, examines, and handles various things which she rejects. Finally she selects a dry oak-leaf about two inches broad and three in length, and lays it over the canopy, and proceeds to fasten it down all around except at the entrance." Thus anthropomorphized, the otherwise scary, hairy spider becomes a caring mother and a nifty architect. Another spider species accepts building materials that Treat offers her. "She takes a stick from my fingers and places it at the edge of her tube. She works while inside of her burrow, holding the stick with her forelegs until it is arranged to suit her."

Some of her pet subjects she even takes into her house: she carefully digs the spider nests out of the ground of the insect menagerie and transfers them to large candy jars that she places in her living room. Then she plonks some eye-catching houseplant on top of each, "so that my nervous lady friends may admire the plants without being shocked with the knowledge that each of these is the home of a large spider."

EUREKA

Mary Treat may be an endearing eccentric, but by and large she is representative of most nineteenth-century scientists. Two centuries ago, even though societies worldwide were developing an insatiable hunger for knowledge, there was hardly such a thing as professional science. So more and more curious people grew into naturalists, and most of those were essentially amateurs, earning a living in some other way and pursuing natural history as a private passion in their spare time. They were personally wealthy, like Charles Darwin and Alexander von Humboldt. They were doctors, diplomats, or clergymen, like the father of genetics, Gregor Mendel. And some, such as explorer and cofounder of evolutionary theory Alfred Russel Wallace, and even Mary Treat herself, earned a living by freelancing, selling their stories to publishers and their specimens to dealers and collectors. But few, if any, were employed by a university or a research institution as scientists today are.

That also meant they did not have a laboratory to go to. For much of the nineteenth century, naturalists made their discoveries in their homes and gardens or in fields and forests. Mary Treat's insect menagerie and the study where she worked with her microscope are matched by Darwin's workrooms and hothouse in the family's Down House, and the "worm stone" in the garden with which he examined how fast earthworms recycle the soil. Gregor Mendel grew his pea varieties in the vegetable gardens of his monastery in Brno, and Alfred Russel Wallace had a new bamboo lab built in every village in the Malay Archipelago, where he and his butterfly net settled for a while (and which he then defended day and night against the hordes of ants that tried to make off with his specimens).

This does not mean they were loners. Like today's scientists, those hestian Victorian naturalists built and maintained correspondence networks that crisscrossed the globe. The 15,000 letters that Darwin received and sent were his database. He scribbled on them, made notes in the margins in different-colored pencils, even cut them up and kept the cuttings in different folders under different keywords. They were "dissected like specimens, every useful bit of information sucked out of them and then reincarnated in his publications," as the Darwin Correspondence Project's website says.

Mid-nineteenth-century naturalists also traveled, visited each other, and organized themselves into learned societies, associations, and clubs, often founding and publishing scientific journals that still exist today. John Tresch, a historian of science at the University of London, says about this period in the United States, "There is a huge flourishing of interest in science from everybody; . . . people going from town to town giving popular lectures. . . . They had experiments, displays, demonstrations. . . . There is a real general thirst for knowledge."

One person who epitomizes this period is the author and poet Edgar Allen Poe. Born in 1809, the same year as Darwin, he is today remembered mostly for his scary tales and for the poem "The Raven." But in his short and tragic life (he suffered rejection, poverty, starvation, slander, gambling, and substance addictions, and died in the gutter), Poe also was a mover and shaker in the American scientific world of his day. He wrote a hugely popular book about shell collecting and classification; he published columns on science and technology in *Burton's Magazine*; and he wrote a 40,000-word treatise on cosmology, *Eureka*, which he considered his greatest work. Tresch, who wrote the book *The Reason for the Darkness of the Night*, about how this little-known aspect of Poe is key to his whole oeuvre, says, "He evaluated all the new inventions and all the new discoveries on the basis of his own experience, having worked in the army as an artificer, someone who makes telescopes and makes munitions, . . . and his voracious reading. So, he is really part of that explosion of popular science."

But a second thread in Tresch's book is how Poe was eventually also instrumental in bringing an end to this era of unbridled growth of amateur science. For in its wake also came a lot of charlatans. Psychics and mystics and quacks similarly went on the lecture trail and gathered large followings; pseudoscience advanced in the slipstream of the true quest for knowledge; fakes, frauds, and forgery were also lapped up by the more gullible naturalists as they sought to quench that thirst for knowledge.

On the one hand, Poe bought into this trend by starting a few hoaxes of his own: he published a story in the *Sun* about a transatlantic flight by hot-air balloon (it never happened), and his well-known short story, "The Facts in the Case of M. Valdemar," in which a dying man is hypnotized and thus kept conscious for seven months while physically dead, was originally published as a piece of true science reporting; only later did Poe admit it was fiction. On the other hand, in his newspaper articles he also exposed hoaxes and frauds, and argued for a national framework for science. In this way, he helped gain traction for attempts from well-known scientists to regulate all this new knowledge, to establish national scientific clearinghouses that were to separate the intellectual grain from the chaff—initiatives that eventually led to the founding of both the American Association for the Advancement of Science and the Smithsonian Institution in the late 1840s.

In other countries around the world, the natural sciences were consolidated in a similar way throughout the second half of the nineteenth century. Large national natural history museums were either founded or taken much more seriously, with lots of public money flowing to them and the public displays becoming showcases for huge research institutes behind the scenes, housing collections of millions of specimens. Universities and private institutions set up large biology research facilities dedicated to understanding heredity, biochemistry, physiology, and the inner workings of the cell, either for pure science or for applications in agriculture or medicine. The German plant physiologist Julius von Sachs, who had been building and running ever-larger and better-equipped university-based laboratories since the middle of the nineteenth century, clearly felt that the days of the private naturalist were numbered. "Amateurish country-house experiments" is how he characterized the botanical studies of Darwin.

Von Sachs's was a sentiment that presaged things to come. By the early twentieth century, people like Mary Treat and Charles Darwin, had they still been alive, would have been anachronisms. In the areas of science where nineteenth-century naturalists had excelled, it was no longer possible to contribute in any meaningful way as an eccentric with a couple of captive spiders in your living room and a worm stone in the backyard. To make scientific progress you needed a large laboratory with expensive equipment, lab assistants, secretaries, and students. You required a well-stocked library with subscriptions to all the major international scientific journals. You needed sizable grants to buy the purest of chemicals and the best of glassware and to travel to international conferences. And more and more professional scientists began to look down on the "amateur" in dismissive Von Sachsian fashion.

2

ROCK VACATIONS

The colors are washed out in the way only 1970s Polaroids fade. The long raincoat of my father, my own jacket with the wooden buttons, my corduroy trousers, the rock debris around us, even the fields on the hilltop in the background—all are the same shade of pale ocher. The photo, which I keep in a drawer of my desk, must have been taken by my mother, and since I look seven or eight years old in it, it would have been in the summer of 1972 or 1973. Somewhere in England or Wales, I guess, judging by the fields in the background, demarcated by hedges and stone walls in typically British fashion. The scene: my father, stooped in that way of his, with his thick-rimmed glasses pushed up onto his forehead and his hair combed backward, as he had always combed it since the 1930s, inspecting a large rock that I, waiting for his verdict, hold up for him. The picture is not quite sharp, but I think I see gray and golden yellow crystals. Perhaps a nice piece of galenite with pyrite.

My father was a mineral collector, and there are many photos like this one in our family albums. Every summer, our small family would go on a "rock vacation." Not too far away, because my father did not like traveling, but since the low-lying Netherlands have no natural rocks to speak of, we had to cross at least one border into Belgium, Germany, or England to get to prime collecting sites. He would harvest the destinations from articles in the magazine of the geological club that he was a member of, and we would spend entire holidays in deserted quarries and mines, eating my mother's sandwiches and cracking open rocks in the shade of our parked orange Ford Escort.

My dad was not just a collector. He also developed tools for his hobby, building Geiger counters to test minerals for radioactivity or dismantling old washing machines in the garage and turning the motors into rock saws or polishers, and then publishing step-by-step how-to guides for fellow subscribers to that same club magazine. Once, he developed a method to cut strips from leftover bathroom tiles, melt, in a hot flame, a pearl of borax on the tip of those strips with some mineral extract in it, and then use a color chart to identify the type of mineral. He published that invention, too, with clear pen drawings and black-and-white photographs to help the novice adopt his method. Another time, with friends from the geological club, he devised a system to identify minerals using punch cards with minerals' physical characteristics (hardness, crystal structure, color, gloss, and so on). If the cards were lined up correctly, he could tell which mineral he had by using one of my mum's knitting needles to probe the number on the card where punched holes went all the way through.

As a child, I was mostly exposed to his mineralogy hobby, but there were and had been many more love affairs with other branches of science and technology in his past. He had played around with optics and chemistry, had collected fossils and gone deep into electronics to build radios, amplifiers, and transmitters. I still have the diary he kept in his late teens in the years immediately after World War II. In an old-style double-entry ledger he wrote in pencil about the slow return of luxury items and food in the shops, his dancing lessons, his job as a painter of advertisement signboards, courting a girl he liked, and shows he visited. But in the back is a very different section, headed, in pencil longhand, "Notes results of experiments etc." One page is about tests with "sound amplification by the 'transitor'" (the transistor was invented in 1947), with wiring schemes and ideas for applications. Another page lists recipes he tested to coat glass with a layer of silver.

Yet another is about physics experiments he did on his own body. "If both my eyes are in the light, and one eye is being shut, then the pupil of the other eye also widens." And: "To observe irregularities in the liquid of the eye, one looks through a very small hole (1/10 mm) in black paper. On the retina, a sharp image is projected of the inner body of the eye. Interesting things become visible, which cannot be on the front of the eye, because they do not change place when one blinks. However, they do make floating movements if one shakes the eye. Also, the objects are not on the retina itself, because they change place when the black paper is moved a little." And there is a little pencil drawing of the cells and detached fibers that my father saw, aged twenty or so, holding a piece of paper in front of his eyes by the window of his small room, in the attic of Segherstraat 11 in The Hague.

He kept up this sense of fun in science and technology throughout his working life (he worked as a graphic designer for a foundation tasked with assisting the rebuilding of the bombed-out city of Rotterdam). Later in his life, after his early retirement around 1980, when the first personal computers became available, he launched into computer science with just as much gusto as he had into building (illegal) crystal radio receivers during the German occupation. I recall how he would spend days on end at our dining table building printers and scanners long before these were available from any shop, and writing software to draw mineral crystal structures or to replace his punch-card system. Even in the weeks before he died (from pancreatic cancer, aged only sixty-four) he was writing code to print address labels for the funeral cards from his Sinclair Spectrum PC.

Despite his technical and scientific brain, he had only vocational school training and, when he worked as an advertisement signboard painter, took evening classes in decorative art at the art academy. He never went to university—it was not something people from his background did. He came from humble beginnings in one of the lesser neighborhoods of The Hague, where his father, my grandfather (who died before I was born), held a series of jobs as candy baker, musician and explicator for silent movies at a movie theater, and concierge at a school. But his father had always had a warm interest in science and had sent him to a small neighborhood primary school run by a husband-and-wife teacher couple who had a keen eye for any pupil with an inquisitive nature. They would take my father to the Schoolmuseum (a kind of "museum of everything" that was frequented by all school classes in The Hague; it still exists under the name of Museon) and stimulate his investigative nature whenever they could. What I mean to say is that my father was a citizen scientist before that word even existed.

And although I did go to university, and was pretty much the first in both my father's and my mother's families to do so, it was not there that I learned to be a scientist. I learned it from him. From a very early age, my father involved me in his scientific and technical exploits. When I was six, he gave me a box of lenses, helped me build a telescope, and turned me into an astronomer for the better part of that year. By age seven, I was his apprentice fossil and rock collector. A year later, I had forgotten about stars and gems and instead was guided by him into the magical world of chemistry. He hand-wrote a book of chemistry experiments for me and helped me turn my bedroom into a chemistry lab. He made me a spirit burner and test tube rack, got me a bag of flasks, beakers, and other glasswork (some dating back to his time as an army nurse) and a starter kit of chemicals. The rest I bought from my pocket money from the local pharmacy—pharmacists at the time would still sell copper sulfate, sulfuric acid, or sodium permanganate to any excited eight-year-old with a good story.

Later in my precocious youth, I would find my own paths into temporary intellectual fads. For one summer, I was an archaeologist, digging up seventeenth-century clay pipes and Roman coins from the fields around our town until I had no more place to put them. Next I turned to electronics, tearing down old radio and TV sets I found in the trash and reusing the components for doorbells, amps, light organs, and mini-synthesizers (none of which I actually needed; they were just so much fun to put together). Then, aged nine or so, I landed in nature study—and remained stuck in it for the rest of my life.

By the time I went to university to study biology, I was a card-carrying amateur naturalist. Expecting to attend lectures with an auditorium full of like-minded classmates, I was disillusioned that almost none of the 120 or so students in my year had a similar background in birdwatching, bug pinning, and botanizing. Most had only a vague interest in biology and had chosen the program as a last resort, other sciences being considered too hard and the arts too soft—either that or they were in a holding pattern while waiting for a place in the medical program to come up. When I told my father about my disinterested classmates he was horrified that anyone would approach a science education, a privilege that to him had been out of reach, so casually.

As I slowly rose in the ranks of professional biological science, I became embedded into a world that was divorced in many other ways, too, from the world of amateur science that I had grown up in and that my father still lived in. In the 1980s, the most important discoveries were published in such print journals as *Nature*, the latest issues of which I leafed through in the university library. Such journals were way too expensive for amateurs to subscribe to, and they were nearly unreadable because they were written in dense scientific jargon, today even more so than a few decades ago. Any calculation more complicated than I could do by hand or on a pocket calculator needed to be written in computer code and sent from the terminal in our office to the big mainframe computer, a humming and hissing thing the size of a minivan housed in a special room next to the botanic garden. (And I could easily take a long lunch break while the computer chewed on my job.)

As a graduate student and then as a postdoctoral fellow in the 1990s I worked in laboratories with expensive equipment for electron microscopy, and for studying proteins and DNA using radioactivity and very toxic gels. Similar labs collaborated in the international Human Genome Project, set up in 1990, the same year that I began my doctorate program, and aiming to read and study all the DNA in the human chromosomes over a fifteen-year period at a cost of 3 billion U.S. dollars. The sheer scale of the expenditure and other resources meant that this kind of science was not within the reach of amateur naturalists like my dad, with their apparatus made from household items and no access to such vast knowledge or computing power.

But things were about to change. That same monolithic computer was connected to something called the internet, and this allowed me to enter instructions into a command line that told the computer to send so-called electronic mail, or email, to a colleague of mine in Austin, Texas. I recall my excitement as the reply to my first email, sent sometime in 1991, rolled out as a string of shimmering green letters on a black monitor screen. Transistors, already a fraction of the size of those my father had been experimenting with at the kitchen table, continued to shrink, speeding up computer processing time and other research hardware. And in *New Scientist*, the popular science magazine that I subscribed to, there was often talk about something called cellular telephones.

The winds of change that fluttered the pages of the typewritten papers I used to hand in as a student have swelled to a veritable storm known as "open science." And open science, a revolution of scientific emancipation that we will discuss in the next chapter, somewhat unexpectedly has brought science full circle. The Mary Treat of today has an entire twentieth-century lab inventory in her smartphone, and the rest she can buy on eBay. She can find any scientific paper she needs at the push of a button and can meet like-minded colleagues and exchange ideas on internet forums. And for any study that needs equipment she does not own, she can sign up at a community open lab. All this has meant that, once again, the "country-house" amateur naturalist sits at science's high table.

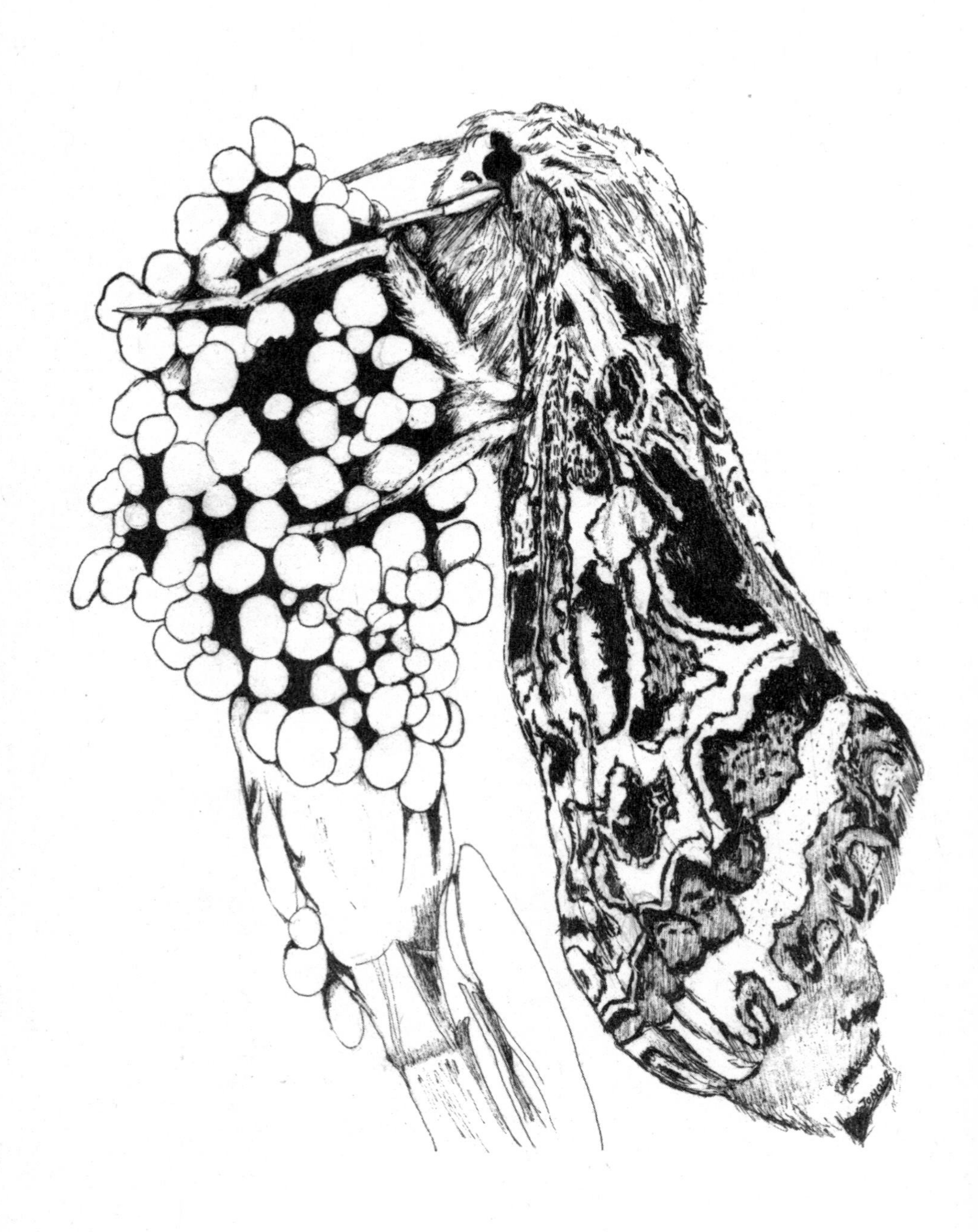

3

DARWIN @HOME

Even though we both speak Dutch, I am having trouble following Patje Debeuf. I come from Rotterdam, he from Ypres, in the deep southwest of Flanders in Belgium, close to the French border. It is only 150 kilometers as the crow flies, but time and all the rivers in between have caused dialects in these parts of the low countries to diverge so much that I catch only one word in three. It also does not help that Patje (short for Patrick) speaks fast and excitedly, for he is over the moon about the rare moth he has just photographed. Blocking out the bright sun with his hand, he shows me the miniature screen of his camera, equipped with a macro photo lens and a huge conical diffuser, and together we peer at the photo he took. It is a *Pharmacis fusconebulosa*, the map-winged swift, he says. The species is difficult to recognize, and though Patje is an expert bird and mushroom watcher, he has only recently started with insects. That he nevertheless is sure of a correct identification is thanks to an application on his smartphone.

ObsIdentify is an artificial intelligence (AI)–driven app that can recognize a good part of the European flora and fauna. Larger animals, but also many butterflies and moths, plants, mushrooms, and even a decent number of bugs, beetles, and grasshoppers, can be tackled by its algorithm. The program has been trained by feeding it with hundreds of thousands of pictures of thousands of species, taken by users of the citizen science platform Observation International (we will meet it again in later chapters), where amateur naturalists log their observations. From such an abundance of images, the algorithm learns to recognize shapes and patterns in pretty much the same way that a human expert would. So, when Patje photographed that to him unknown moth, he fed the image (chunky and hairy, with a delicate pattern of brown and white markings on its wings) into the app, which immediately pronounced, with as much bravado as a seasoned moth expert would, "image recognition predicts Map-winged Swift—*Pharmacis fusconebulosa* with 100.0 percent probability."

Truth be told, when I first started working with ObsIdentify I had my doubts, but soon, if grudgingly, had to admit its superiority and magnificence. Grudgingly because this bit of stupid computer code was able to do snap determinations of species that had taken me years to learn to distinguish. Superiority because I was blown away by its accuracy. Even an out-of-focus, washed-out picture of an insect taken at a weird angle can be identified by the app flawlessly. Perhaps I would be able to reach a similar level of accuracy for my pet organisms, the beetles, and maybe also for land snails and a few families of birds and plants, but that is where my hard-earned expertise would end. Nor could I—or anybody else, for that matter—ever single-handedly and equitably recognize species of moth, plant, butterfly, amphibian, and mushroom, the whole shebang, as ObsIdentify can. The app is no less than the digital equivalent of an unbelievably skilled polymath naturalist who has spent a lifetime looking at flora and fauna, buying expensive identification manuals, leafing through determination keys, peering through microscopes, and practicing, practicing, practicing. And while ObsIdentify works very well for parts of Europe but is yet to be trained for other continents, other image recognition apps, such as Pl@ntNet (for plants), Merlin (birds), and Seek (for flora, fauna, and fungi), work nearly as accurately at a global level. Some, like Merlin and BirdNet, can even be used to record and identify sounds (the algorithm for this is trained in a similar way as the image ones are: sounds are first converted to sonograms, and the images are then treated just like photos). And binoculars manufacturers are even beginning to integrate image recognition into their optics: when you point Swarovski's new AXVisio binoculars at a bird, the identification will appear inside your field of vision.

At the moment I write this, these apps are still limited by the amount of available training material. On the citizen science platform iNaturalist, the well-known and loved, globally abundant seven-spot ladybird beetle, *Coccinella septempunctata,* is represented by a whopping 95,000 "research-grade" (that is, considered reliably identified by a two-thirds majority of users) photos. Few of them have been taken with a high-end camera like the one Patje wields; most are simply smartphone snapshots. They come from all over the world, including such unlikely localities as Reykjavik, the Indian Ocean island of Réunion, and the driveway of the Library of Congress in Washington, D.C. They include color aberrations that are entirely yellow or with the black spots triple the usual size; are out of focus, motion-blurry, or black-and-white; in close-up or just a dot on a whitewashed wall; taken from above, the front, the rear, or the bottom; perched atop a fingernail or pinned in a collection; mating, in hibernation among a hundred other ladybirds, with wings extended or belly-up dead. With such an abundance of reference material, no new image of a seven-spot ladybird will be so unexpected as to be unidentifiable by iNaturalist's app Seek.

But another, globally equally common beetle, known only by its scientific name *Cartodere nodifer*, is, although no less cute, much less popular among those who photograph insects: it happens to be smaller and browner, and lives in less attractive places (mostly compost) than ladybird beetles. So, on iNaturalist you can find only a miserable 190 research-grade photos of it—not enough to train the AI algorithm, which does not do any better than saying in a small voice, with only a shimmer of its usual confidence, "We're pretty sure this is in the genus *Cartodere*." However, as more and more photos are uploaded and more and more experts join iNaturalist and the other platforms and help identify photos, it is only a matter of time before the identification of obscurer insects like *Cartodere nodifer* and virtually any other species of animal, plant, or mushroom can be handled by an app on your phone. Not only are these apps going to be more rapid, more broadly specialized, and more accurate than any flesh-and-blood naturalist, they will also help turn people like Patje Debeuf into experts.

And that is where their real power lies, as I realize when Patje Debeuf tells me his life story over a beer. He started out as a birder when he was a six-year-old, in the late 1960s. And birder, in those days and in that part of Belgium, meant not bird-watcher but bird-*catcher*. The catching of small songbirds with glue, nets, and traps (for food, and to be sold in the caged-bird trade) was immensely popular in Belgium, as it sadly still is in some Mediterranean countries. Even in the early 1970s you could still buy wild-caught birds at the weekly bird market in the town square of Antwerp, where they were kept in noisy stacks of tiny, dank, smelly cages. In his hometown, says Patje, there were a few professional bird-catchers that he and other children of the village often trailed to lend a hand. It is not something he is proud of, but back then it was a normal part of rural life, and it did provide him with a lifetime supply of ornithological know-how.

When bird-catching was declared illegal in Belgium and birders with binoculars began replacing those with nets, Patje also turned into a bird-watcher and bird conservationist. Not that his professional life had anything to do with his hobby. The day jobs he had held since he left school at the age of sixteen were packing and carrying meat and driving the local meat wholesaler's truck. He shows me a picture of himself in his younger days, posing behind his truck, dressed in blood-stained butcher's white, with what looks like half a cow on his shoulder. "That is how I destroyed my spine," he says ruefully. Meanwhile, even though his workmates could not quite get his nature passion, he grew into one of the best and most experienced birders of the region.

After an early retirement because of his back problems, his community scientist work intensified. He joined (and later became a board member of) his local nature club, got interested in orchids and mushrooms, and started training as an official volunteer nature guide. He also joined the excursions of the insect working group. "I began to be fascinated with insects," he says. But the tens of thousands of species of insect, even in a small country like Belgium, and even to a seasoned field naturalist like him, were daunting. "It used to be such a hassle to identify them; I had to borrow lots of books and spend a lot of time on it." But when ObsIdentify came around that all changed. Quick, AI-based identifications allow him to identify his sightings much more swiftly. And once he has a scientific name he can upload his photos and records to Observation International and begin investigating what is known about a species. Where does it occur? What does it feed on? Is it rare or threatened? Moreover, he will memorize the name and recognize it when he sees it again.

The new tools also help him to inspire young people. Once he finishes his nature guide training, Patje will start a series of lessons for primary schoolchildren. They can join him on his excursions, learn to handle a butterfly net and a camera and use the apps on their smartphones to get to know and understand insects. Just as he, I imagine, sixty years ago was spellbound by the bird-catchers and their lore and tricks of the trade.

A LAB IN THE PANTRY

AI is beginning to be an important equalizer among professional and amateur biologists. No longer does one need access to an entire library, whether physical or digital, of identification keys, field guides, and manuals. No longer does one need to learn the technical terminology to distinguish this species from that, or spend years training to recognize the subtle differences among multitudes of near-identical species of organism. A decent photograph and a well-trained image recognition app do the trick, or soon will.

But perhaps I am being too optimistic. Perhaps the thousands of different slime molds, mites, nematode worms, and parasitic wasps will never be sufficiently known for an algorithm to encapsulate and democratize the knowledge about them. And certainly, organisms that have virtually no physical distinguishing marks to speak of will be out of reach for any image recognition application: many species of fungi, protozoa, and bacteria are simply too similar in appearance to ever tell apart visually. The differences among them lie in invisibilities of their physiology, the way they process nutrients and find their favorite corner in their chemical and physical microhabitats. Only the structure of their DNA betrays their identity. And DNA can only be studied in a professional laboratory, right? Molecular genetics is not something an amateur naturalist can master at the kitchen table—or is it?

Meet today's citizen science genetics sleuths and their kitchen tables.

One of those kitchen tables belongs to Sigrid Jakob. Originally from Germany, she now lives in Brooklyn, New York, has an arts degree, is fifty-something and the mother of a teenager, and earns a living as a freelance strategist for large companies. She emphatically is not at all, as she says, "your typical bio-hacker type." Yet somehow she blundered into the world of mushroom DNA. Around 2015, "God knows how," she developed an interest in fungi. She joined a local mushroom club, got hooked on identifying species ("it gives you that little dopamine hit—it's very addictive"), and would go out every weekend to look at fungi. But, she says, she got "really frustrated when you find something that's not in the field guides; you ask people, what is it? and nobody can give you an answer."

What Jakob describes is one of those common cases where species differ by other things than external appearance, and image recognition apps, even detailed scrutiny under a microscope, cannot reveal the differences. Biologists call them sibling species, and they abound in nature. Usually there *are* important differences between them, but these lie not or very little in the way they look and much more in how they deal with their environment. Two mushroom species could be nearly indistinguishable, but one has its mycelial threads intertwined with the roots of oak, for example, the other only with those of poplar. It is a crucial difference and one that the entire life cycle of the fungus depends on, but completely invisible to the human eye unless those eyes can penetrate into the very DNA in which those species differences are encoded. What Sigrid Jakob realized, in her mycologist frustration, is that she needed to grow precisely a pair of eyes like that.

Until just a few years ago, using DNA and biotechnological tools to recognize species and to work out how they are related to one another was an enterprise firmly inside the realm of academic science. To sequence even a short piece of DNA was a time-consuming and above all very expensive undertaking that required equipment costing hundreds of thousands of dollars apiece.

But things have changed since the 2000s. Technology has become miniaturized, cheaper, faster, and easier to use. The human genome is a case in point. When in 2003 the first sequence of the human genome was generated, the whole international enterprise had cost hundreds of millions of dollars and taken over a decade. Today, as I write this, you can have a human genome sequenced by a small lab in one day for a few hundred dollars, and perhaps by the time this book is printed and read by you that price will have been slashed even further. The incorporation of DNA practicals into science curriculums in schools has led to the development of small, educational, affordable biotech equipment that can be bought off the shelf. And the boom-and-bust cycles of small biotech startup companies has flooded the second-hand market with cheap used lab equipment. In other words, the time was right for even a curious, confident arts major with an interest in 'shrooms and a few hundred dollars to spare to build a small DNA lab on her kitchen table.

Jakob's whole lab fits in two shoeboxes, and she can quickly unpack it on her kitchen counter, use it, and make it disappear again before the kitchen counter has to be used for preparing dinner—"it's very family friendly." To get there, though, was a bit of a struggle. She waded through piles of online tutorials and protocols, and not everybody on the online forums was equally supportive. "It seemed super intimidating. I do not have a scientific background at all." Another problem was that until she discovered an online shop for the bio-hacker community, she had a hard time sourcing the chemicals she needed. The "official" chemical supply companies would stop her in her tracks with questions like "What is your business address?" and "Can I speak to your finance department?" But in the end, she managed to set herself up with a working DNA lab.

Working, that is, to do something called "DNA barcoding" on her mushrooms. DNA barcoding has its origin in the 1980s and 1990s when biologists discovered that every living organism has parts in its genome that have just the right amount of variation to use it as a handy way to tell two species apart. Take the famous 2 percent difference in the DNA of chimps and humans, for example. A 2 percent difference means that, if you were to align the entire genome of a chimp with that of a human, on average, two out of every one hundred DNA "letters" would be different. The "on average" here is important. There are some genes that are so crucial to the functioning of any organism that genetic mutations are simply not allowed, and any that occur will have been removed by natural selection. Those "conserved" genes will have near-identical DNA sequences even between a fruitfly and a fungus, and certainly between a chimpanzee and a human; thus, such genes cannot be used to tell whether you are dealing with human or chimpanzee DNA.

Conversely, other genes have too much variation: they may have lost their biological function and mutate at random with such abandon that every human or every individual chimp will possess a unique sequence. Such "hypervariable" genes are great for DNA fingerprinting in forensics to pinpoint a killer but useless to tell entire species apart. But then there are some "Goldilocks" parts of the genome that are just right: they are sufficiently similar within a species to genetically tell whether samples come from the same species, and uniquely different between species to make sure you are dealing with different species. In other words, such fragments of the genome can be used like a barcode to quickly scan to identify a species—provided you have a catalogue with reference sequences for each of the potential species.

For fungi, the gene of choice is called the "internal transcribed spacer," or *ITS* for short. When Sigrid Jakob finds a mushroom that she cannot place, she first snips off a piece the size of "half a grain of rice" and crushes it, on her kitchen counter, in a small plastic test tube with lots of chemicals to squeeze out all the DNA from the fungal cells. Then she mixes the DNA extract with several ingredients (which she keeps in a bag in her kitchen freezer, next to the frozen peas), namely, loose DNA building blocks (the four "letters"), an enzyme that can copy DNA, and two primers, which are short runts of DNA that latch on to either side of the *ITS* gene in her DNA sample. Then she sticks the test tube into her PCR machine. PCR, which stands for "polymerase chain reaction," sounds fancy, but a PCR machine is essentially simply a gizmo that can very quickly cool down and heat up mini-test tubes. With each cycle of heating and cooling, the *ITS* gene doubles, leading to an overwhelming amount of *ITS* DNA that, after a couple of hours of humming away on her coffee table, is enough to be read by a sequencing machine.

Jakob does not own a sequencing machine (though some kitchen table amateur geneticists do because they, too, are getting cheaper and smaller all the time), so she sends the PCR-treated sample off to a biotech company that, within a few days, emails her the string of letters that is the DNA sequence of the *ITS* gene in her mushroom ("A's, C's, G's, and T's, hundreds of them, ad nauseam," she says). Finally, the moment of truth: she goes online to a database that holds reference DNA sequences—GenBank, for example, which harbors the world's entire knowledge of DNA of hundreds of thousands of organisms. Now, you might think that that is a huge database, but since DNA data are just strings of letters, which take up very little memory space, even the ten trillion DNA letters that GenBank contains today amount to a computer file that fits onto a flash drive—which means you could even use it offline.

In GenBank, she pastes her DNA sequence into a search window, and up come the best matches. If there is a good—say, 98 percent or higher—match with a known fungus, then she has made a positive identification. But sometimes even GenBank does not come up with a match. In that case, as has happened many times for the more than one thousand sequences she has generated in her home lab, it may turn out she has discovered an entirely new species. That is not that surprising, she says, since scientists estimate there to be around 3.8 million species of fungi in existence, but only 150,000 have so far been named and catalogued. "The chances of finding new species are shockingly high," she says.

It is ironic, really, says Jakob as she reminisces about her life and what choices got her to where she is now—newly elected president of the New York Mycological (fungi aficionado) Society and studying fungal genetics from her home in Brooklyn whenever she has some spare time. Growing up in a small village full of outdoor folk in southern Germany, as a teenager she hated all of that, the endless talk about rural topics. She says, "I was punk, I was goth, I was shy, I was weird, I dropped out of high school. Working in a pants factory got old really fast, so I went on a German game show, made some money and ran away to London—note on pillow, early morning train, that kind of departure." And now, after having followed a tortuous path in the UK and the United States through waitressing, secretarial work, university and art academy, and into the corporate world, she finds herself reveling in the same sort of outdoor life that she eloped from nearly forty years earlier, taking her teenage daughter on nature walks and lecturing her about the Latin names of mushrooms and trees: "Now *I* am that boring mum!"

And it is not only her family that gets regular mycological tutoring. She has also been an inspiring force in the New York community of amateur mycologists, many of whom she has helped overcome their fear of DNA. She has given workshops and has helped build kits for DNA barcoding at home, so members of the club that she is the president of can borrow the equipment to try it out on their own kitchen counters.

And from there it is only a small step to real community labs.

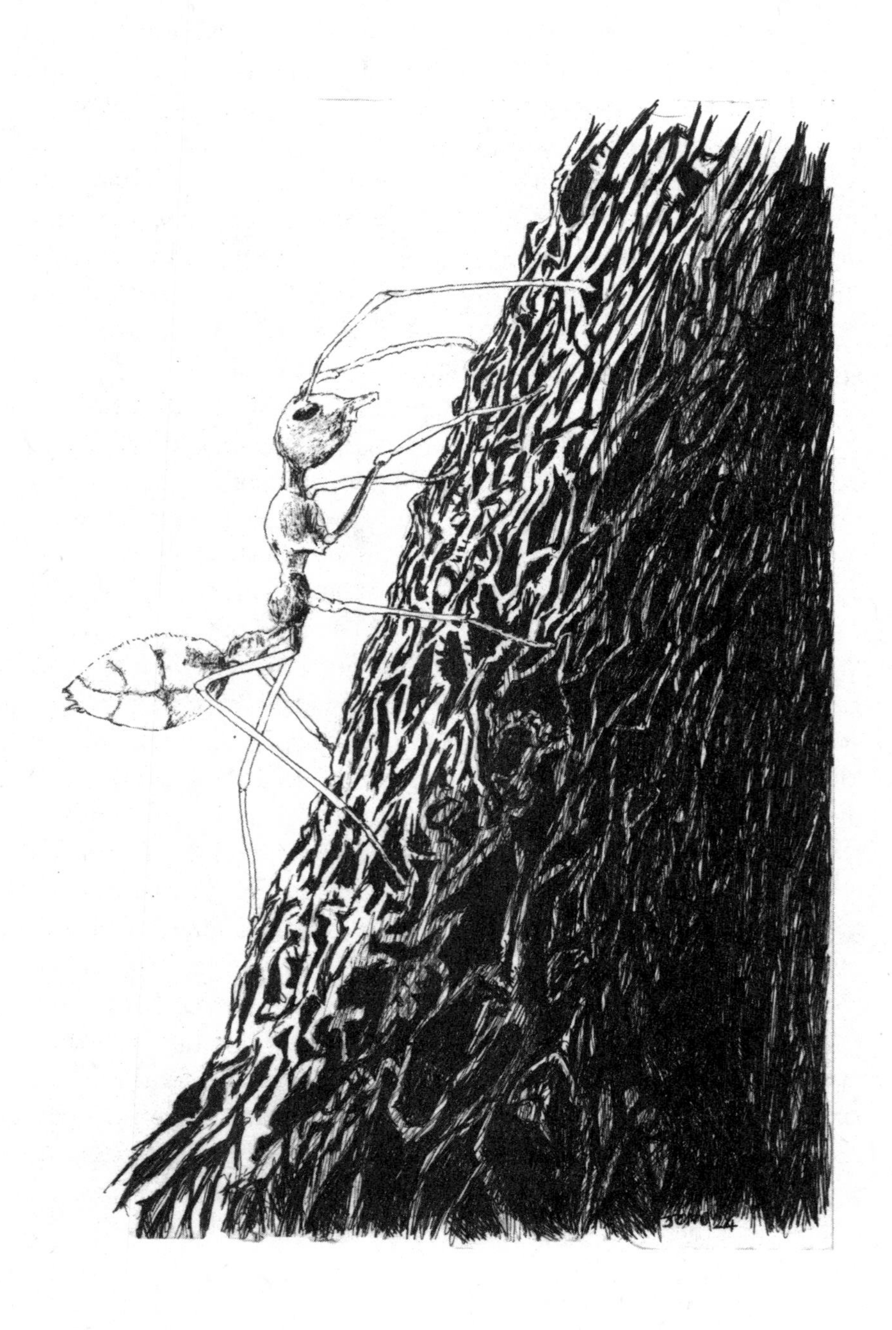

4

LABS FOR EVERYONE

If you want to be a community scientist but you do not want to go it alone (after all, a loner community scientist is a bit of a contradiction in terms), you could join a community lab. Traditionally, natural history museums and science centers have long supplied space and opportunities for amateurs to use their facilities in off-hours. When I started out as a schoolboy with a naturalist passion, I used to attend the amateurs' days at the Zoological Museum of Amsterdam. One Saturday per month in the fall and winter—in spring and summer everybody would be out in the field—the museum would open its collections and labs to amateur collectors from all over the country, who could use the museum's multimillion-specimen collection to compare their finds against. The tables in the collection space, thick with naphthalene vapor, would be crowded with men and women of all ages, from wide-eyed teenagers like me to centenarians who survived purely because of their natural history passion. We would huddle excitedly around microscopes, hastily finish our sandwiches in the museum canteen, stay as long as we could, and in the evening commute back by tram and train to wherever we came from, oblivious of the pong of naphthalene that emanated from our clothes for the entire trip.

But in recent years, especially for those community naturalists with a more experimental bent, independent, community-run lab spaces have sprung up in many cities all over the world. The website of DIYbio (a contraction of "Do-It-Yourself Biology") lists sixty-three such labs on all continents. True, most support not just naturalists per se but a motley crew of biohackers, tinkerers, amateur informaticians, aspiring biotech startups, and bio-artists, but they will give the novice a warm welcome and help them learn how to use, build, and modify scientific equipment, develop an investigative mindset, and start communal urban naturalist projects.

Members of the Baltimore Underground Science Space (BUGSS), for example, organized themselves around a project to do DNA barcoding of the organisms living in the Baltimore Inner Harbor—or rather *surviving* there, for the urban estuary is heavily polluted, despite recent cleanups and restoration programs. At specific locations in the harbor, the BUGSS members placed biodisks in the water, plastic devices on which, over time, small sea creatures would settle: mussels, barnacles, anemones, snails, and other marine organisms collectively carrying the misnomer of "the fouling community."

Once a month, the community scientists on duty would haul up the biodisks and scrape into a bottle of ethanol all the newly settled animals, then take the sample back to BUGSS, located in a drab brick building in an industrial complex not far from the harbor, for DNA barcoding. Without the need for detailed marine biology skills, the project, called Barcoding the Harbor, could thus monitor the changes in the harbor's biodiversity simply by keeping track of which DNA barcodes participants could squeeze out of their biodisk samples.

These largish community biolabs can help members do things and reach goals that they could not with their own limited means on their kitchen table or in their garden shed. Genspace in Brooklyn, New York, for example, says Ellen Jorgensen, cofounder and president, "is filled with equipment that is either bought off eBay or donated or on loan." They have a spectrophotometer, an incubator-shaker, cell culture facilities, gel electrophoresis equipment, PCR machines, microscopes, and shelves replete with smaller equipment, vials, glassware, and consumables. In fact, it looks pretty much like a regular cluttered university lab and has a large staff of managers and instructors and, presumably, rules, policies, and housekeeping regulations. That is great, because it means that the community of users can get all the support and material they need, and that there is a free flow of ideas among them. But does it not also run the risk of losing the bottom-upness, independence, and freedom that are so crucial for community science?

I put this question to Jan-Maarten Luursema, an artist, tinkerer, and DIY biology enthusiast based at Radboud University in Nijmegen, the

Netherlands. "I don't think so," he replied. "Even large community labs like Genspace are still very much grass-roots." But such labs also feel that other sections of the community in which they are located eye them with suspicion: *What if something evil escapes from those incubators?* So more and more labs are being pro-active in drawing up ethics and safety policies and even sharing these with the authorities and the public preemptively.

Although many community labs have a clear biotech and biomonitoring bent, some hark back to the traditional natural history collecting. The heart of the Nature Lab in Providence, Rhode Island, for example, is a natural history collection of 100,000 specimens, an eclectic assemblage of rocks, fossils, skeletons, shells, microscope slides, varnished fish, pinned insects, and stuffed mammals, displayed in the old cabinet-of-curiosities way, on shelves behind oak-framed glass doors, in rooms with huge puffer fish suspended from the ceiling and deer's antlers mounted on the walls. Anybody can come and draw, study, photograph the specimens, or use them in any other way, even remotely, for many are available online as 3D digital models. There are benches with microscopes and macro photo setups that you can sign up to use, and lots of other shared resources.

Other community labs are networks rather than places. Public Lab, for example, working out of a coworking space in New Orleans, Louisiana, was started by a group of rebels in 2010, during the BP oil spill in the Gulf of Mexico, the largest accidental oil disaster in history. At the time, BP was put in charge of access to the area, which effectively meant a blackout for all objective information about the extent and impacts of the spill. A group of concerned locals, environmental advocates, designers, and social scientists got together and began building their own "community satellites" from balloons, kites, and lightweight digital cameras, which they flew over the affected area, staying under the radar of BP. The more than 100,000 aerial images they took were placed on an open-source platform and stitched together to make real-time maps that found their way to conventional media outlets such as the BBC and the *New York Times*, which were just as starved for reliable information as the local residents. Galvanized by this success, Public Lab continued working as an international social network of community scientists and tinkerers focused on environmental conservation.

FRUGAL TINKERING

For community scientists who still would like to build their own private lab at home, fortunately, things are also getting easier and easier. Not only can you buy cheap, second-hand or off-the-shelf DNA equipment but also good microscopes, and even electron microscopes can be had on eBay or equivalent consumer-to-consumer platforms. For a few thousand dollars, you can buy an actual scanning electron microscope. You would probably need to rebuild the shed to fit it and the bottles of liquid argon gas that come with it in, but then you would be all set up to view your specimens at magnifications of tens of thousands of times.

But not everybody has that sort of money lying around to spend on what to most community scientists essentially is a hobby. In that case, you could start following a movement called "frugal science," which aims to MacGyver cheap, simple, durable devices that make biological research more accessible to more people without an appreciable loss in performance. A good example of this movement is the work by Manu Prakash, professor of bioengineering at Stanford University.

Prakash is best known for his award-winning Foldscope, a functioning high-power microscope costing only a couple of dollars and created from nothing more than an LED light, a battery, a single beadlike lens, and some pieces of cardboard, which get origamied together into a gadget that allows you to view specimens at 140 times magnification—a bit like those tiny hand-held brass microscopes used by the seventeenth-century founder of microscopy, Antonie van Leeuwenhoek. Prakash got the idea for the Foldscope, he says, when he visited a lab in Thailand where there was a microscope that was so expensive that nobody dared touch it for fear of causing several annual salaries' worth of damage. Unlike white elephants like that, the Foldscope now has nearly two million users worldwide, who share their discoveries on an online platform called Microcosmos.

On December 4, 2022, for example, a Foldscope user in Madurai, India, uploaded a beautiful picture of spiral chloroplasts inside a freshwater alga—bright green photosynthetic necklaces helically arranged along the inner cell wall. And a few weeks earlier another user, Lukman Rizkika, from Surakarta, Indonesia, photographed the formidable set of serrated jaws of a termite from his backyard. That they can do so is thanks to a handy magnetic attachment of the Foldscope that allows you to link it to your smartphone camera.

The smartphone itself is a device that can hold an entire lab bench full of science devices. Obviously, you can turn it into a GPS locator, a sound recorder, or a stopwatch, and we have already seen the AI-based apps for identifying and recording species. But you can do so much more with it. You can, for example, go even lower-cost than the Foldscope and create a microscope, even if your smartphone does not have a macro option. If you turn on the selfie camera and carefully place a water droplet precisely on the middle of the selfie camera lens (which is protected by a waterproof screen, so no worries), you have created a microscope with a photo and video camera attached, as you will see when you hold an object very close (a few millimeters) above the droplet. And you can increase the magnification from five times with a large droplet to up to fifty times with the smallest droplets.

Also, smartphones have great computational power, and a lot of sensors for tilt, movement, light intensity, sound, acceleration, and magnetic fields. Many free science apps make use of this. For example, the app Canopeo measures canopy cover, the percentage of the sky that is covered by foliage—an essential metric in any ecological field study. Another app, LeafByte, calculates from a photo of a leaf how much of it has been munched away by a caterpillar or another herbivore. And with Arboreal you can measure the height and crown size of a tree. The possibilities are almost endless.

With add-ons you can enhance the naturalist power of your smartphone even further. Want to peek inside a bumblebee nest or a gopher burrow? Just snake in one of those endoscope smartphone attachments, the cheapest of which costs only a couple of dollars. With the iSpex attachment, developed by the astronomer Frans Snik, you can measure the amount and types of aerosols in the air. And the company Oxford Nanopore says it is developing the SmidgION, a smartphone add-on that will have an entire DNA lab inside that can analyze and read DNA samples in real time, pretty much the way Spock's tricorder device did on the *Star Trek* TV show. At the risk of being overly optimistic, I'll hazard the prediction that within five or ten years after this book is published, tricorder-like mobile devices that allow you to scan DNA barcodes whenever and wherever you want will no longer be science fiction.

Of course, a community naturalist's ambitions need not be limited by what can be purchased from others, no matter how cheaply. Equipment can also be self-built, and designed and refined to suit the scientific purpose. In other words, low-cost tinkering at home can suffice.

My friend Andrew Quitmeyer is a person who embodies the concept of low-cost tinkering, and he does so in a totally new way. Having coined, developed, and personified (as lead protagonist in the 2017 Discovery Channel series *Hacking the Wild*) the field of "digital naturalism," he started out as a filmmaking and engineering student at the University of Illinois. Being interested in science, he signed up for as many science electives on the side as he could. One day during his graduate studies, while he was working on a robotics project in a lab that studied ant behavior, his entomologist collaborators invited him on a field trip. "It was so fun and interesting I had a sort of existential meltdown." He recalls thinking, "Oh no! I screwed up. I could have been a field biologist all this time? Like, that's a job?." Realizing that it was too late to completely switch to biology, he did the next best thing and embarked on a multidisciplinary (he prefers to call it "antidisciplinary" himself) mashup of unconventional technology and animal behavior.

Unconventional, because Quitmeyer goes against the grain of much technology-driven field biology, which tends to rely on expensive, moonshot-type equipment. "The whole reason that I call my work 'digital naturalism' is that I am building off of this more romantic side of this ecological science that was developed in the 1800s and 1900s," he says. Digital naturalism's essence is, he explains, an attitude to not ask, "What technology can we shove out there?" but rather, "What are the values of field biology and how can tech support that?" To reach that goal, he organizes what he calls "hiking hacks." He takes teams of technologists, artists, craftspersons, biologists, and naturalists from a wide variety of ages and backgrounds on camping trips where they can bring only the simplest of tools and hardware, and creates an atmosphere of full immersion in the natural world. Cut off from the outside, and with so much human talent, experience, creativity, playfulness, and intellect concentrated under a single tarpaulin in the middle of a forest, naturalist questions and digital tools to answer them develop hand in hand.

"We set up a jungle maker space, build worktables out of woven sticks and vines, and use butane-powered soldering irons to try to make all the crazy devices we can while we are out there." Or, while using the wearable lab that his team member Hannah Perner-Wilson has been developing: "She is making these really gorgeous portable, hacking workspaces; a backpack that completely unzips, and you can hang it on a tree, and it has all of your conductive threads already in little areas you can pull and snip off, your tools organized. . . . It is a really gorgeous idea."

Out of Quitmeyer's digital naturalism has flowed an endless stream of cute, simple, and whacky inventions, some useful, some unapologetically useless. Here is just one example to give you, quite literally, a flavor: the insect traffic taster. The idea, says Quitmeyer, was born out of the wish to monitor the activities of ants in a tree. Ant researchers like to understand how ants forage in the forest. But when a colony of ants explores a tree they are everywhere, workers diffusing inextricably into the complex three-dimensional structure of the forest.

The conventional way to monitor ants would be to just stare at the tree surface and make notes, or perhaps aim a bunch of cameras at the tree and that way try to make sense of what they are doing. But, says Quitmeyer, "your vision is actually quite limiting [as] to what you can take in from the world." So Perner-Wilson and he rigged a rainforest tree with bundles of fiber-optic threads. They first sanded the threads so that the light flashing through them would scatter outward and interfere with ants crossing over them. The changes in light patterns would then send back signals to a central processing device that turned these signals into individual currents.

Now, a more conventional, less playful technologist might display these output signals on a computer screen or a paper readout. Not Quitmeyer. He decided to use one of the tree's own leaves as output terminal by sewing all the thin copper output wires into the leaf. You can then take that leaf into your mouth, press your tongue against it, and use the tongue's "high-resolution spatial imagery" to create a mental picture of the ants roaming around the tree from all the tiny electrical currents that your tongue would be picking up as "tickling sensations."

In a short video he produced, we see Quitmeyer dressed in a raincoat and high up in a cloud forest tree in Madagascar saying, "If I want to know what's happening with my ants, I just go . . ."—and this is where he folds his mouth around the special leaf and starts mumbling about what his tongue is picking up of the ant's movements—"Hmmm. . . . Hm-hmmm!

Wherezum-whombewow. . . . Ahrwittledickwingand. . . . Andzenzhawthing. . . ." (He removes the leaf from his mouth.) "If you got good at it, you just sat there, and kind of meditated with it, you could get a pretty good idea of these waves of traffic and how they move through the varying shapes and contours of this tree."

It is essential, says Quitmeyer, to develop tools smack in the natural situation where you plan to be deploying them. Some people will "come in with a whole bunch of ideas and they'll get all prepared and be set on this one specific idea: 'This is the idea, I'm gonna go into the forest, I'm gonna deploy it, it's gonna be great.' And over and over again, . . . in those first five seconds when you pull out the thing, that you've been thinking about for months, and you're like: 'Oh, wait a second, this might not work.'"

In 2018, in his mid-thirties, Quitmeyer gave up his academic position as a professor at the National University of Singapore. Exasperated by the academic culture of chasing money and citations and credentials, he followed a childhood dream and moved to Panama. "It had always been my 'master plan' that when I'm fifty then I'll quit everything and then maybe I'll have my own field station then I'll go gung-ho and do this." When his frustration in Singapore reached a boiling point, his partner said, "Screw that. Just do this *now*."

In 2019, they moved to Gamboa, Panama, where Quitmeyer had worked with his ant collaborators way back when as a PhD student, and set up their own digital naturalist lab (or DinaLab) in a timber house on the fringe of the Soberanía National Park. They continue to organize hiking hacks there, and collaborate (as "full-time weirdos") with the scientists at the Smithsonian Tropical Research Institute next door.

In one project, for example, they tampered with the software that controls their 3D printer to make it suitable for printing with melted discarded laboratory plastics from the Smithsonian (think of all the single-use pipette tips, vials, and ninety-six-well plates that laboratories spew out in vast quantities all the time). And then they used that plastic to 3D print new lab equipment.

3D printing itself is an essential part of open and frugal science. Instead of buying your equipment and gadgets from a supplier, you can download a 3D digital model that somebody, somewhere, has uploaded to one of the many online libraries that exist for this purpose, feed it into your (or your community lab's) 3D printer, and print your own laboratory centrifuge, microscope, pipette, or forceps.

Over the years, Jan-Maarten Luursema, whom I mentioned earlier in this chapter, has contributed 3D models for printing tube racks, centrifuges, and an "arena" for growing slime molds, to name but a few. He shows me a white plastic paddle wheel that he designed and 3D printed. He clicks it open. and out comes a cylindrical core with holes in it. It is, he says, an automatic DNA sampler. You can put a filter inside the cylinder and then suspend it from a tree, where it spins in the wind, collecting pollen and dust from the air as it does so. After a few days or weeks you can release the filter and extract all the DNA from it, run a DNA barcoding analysis, and obtain a list of all the local species of flora, fauna, and fungi that have left bits of their DNA floating in the air in the area.

Sampling such "environmental DNA," or eDNA, for short, is another new way in which DNA barcoding is democratizing expert knowledge. All organisms shed cells and even loose DNA molecules into the environment. As a valuable carrier of genetic information, DNA is a resilient chemical compound that stays in shape long after it has left the organism it came from. Renegade strands of DNA of nearly all organisms that live in a certain place are literally blowing in the wind (or floating in the water, or buried in the soil), and can be picked up by a filter like Luursema's. With the use of techniques that amplify and read DNA barcodes in bulk, such samples can then yield catalogues of large parts of the biodiversity, in an automated fashion, without the intervention of any expert. Again, the time is near that community scientists can use these techniques to whip up quick biodiversity inventories of greenspaces, which may even reveal species that are there but too elusive to show up on conventional binocular-and-microscope assessments.

In this chapter, we have talked about science equipment: how you find it, make it, use it, play with it, and apply it to new uses. But those tools of science are just that: tools. And tools alone do not make science. They may help you to gather information, but the next step is to organize that information into scientific stories and scenarios, and compare your findings with those of others. And that is where we get to that ocean of information that is the scientific literature.

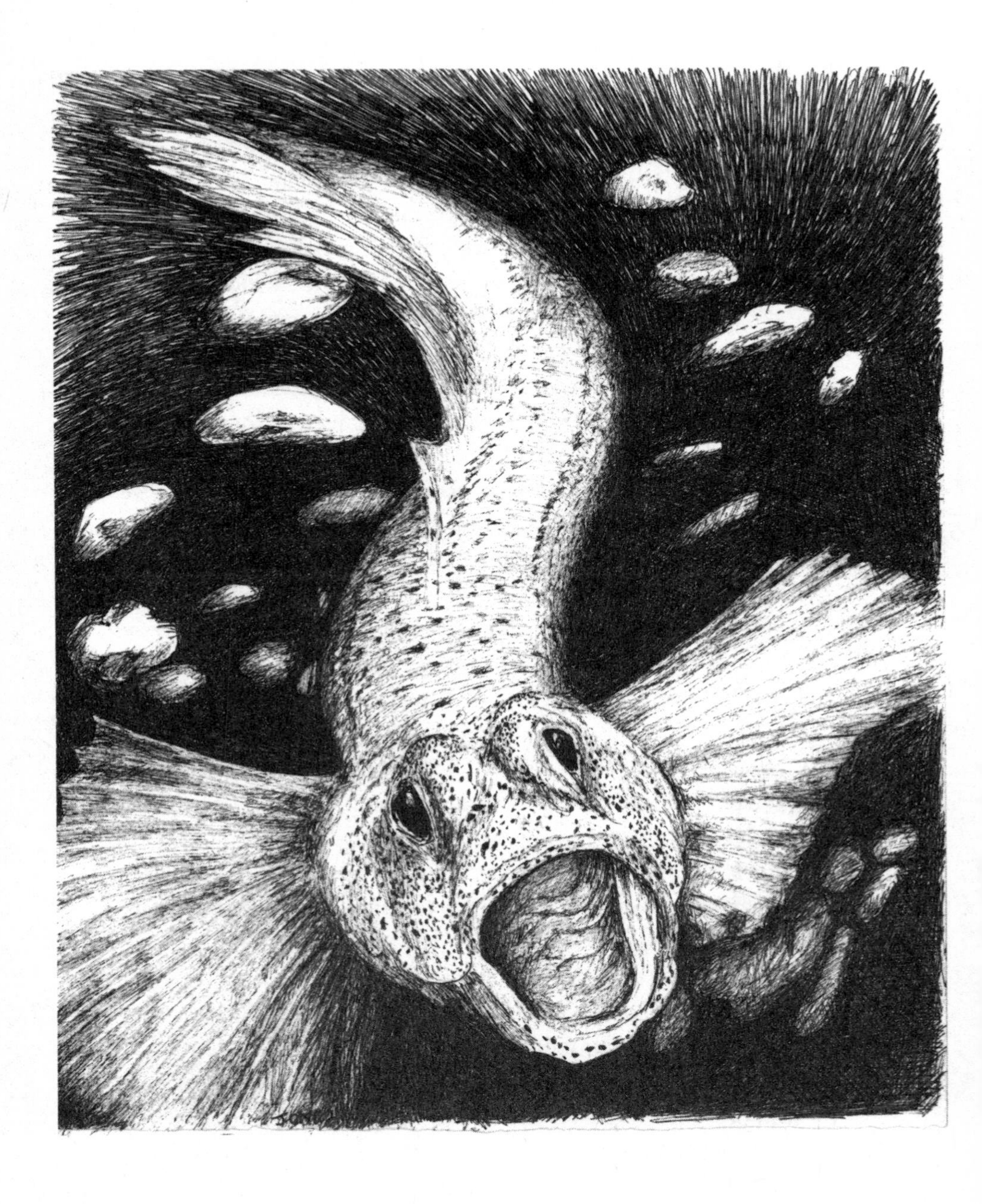

5

VIRTUAL ACADEMIA

In the natural history museum in Leiden we used to get regular visits from Akihito, a friendly community scientist from Japan and an expert on the fish family Gobiidae, or gobies. The reason why Akihito would come so far to study our fish collection is that the Leiden museum holds some very important historical specimens from Japan, so-called holotypes, or universal reference specimens. These were collected and preserved in alcohol by the Dutch zoologist Philipp von Siebold, who, in the 1820s, was the only European permitted to work in Japan, which at that time maintained a strict isolationist foreign policy.

Akihito comes from a Tokyo-based family with a history of community science: his father was interested in crustaceans, and Akihito's younger brother studies birds. As they are fairly wealthy, Akihito and his relatives run their own zoological laboratory at their family home in the old city center of Tokyo. Not only that, they also publish articles in prestigious scientific journals such as *Ichthyological Research* or *Gene*, using—and this is unusual—their home address as their only affiliation.

Now, Akihito probably gets away with this because his home address happens to be The Imperial Residence, 1–1 Chiyoda, Chiyoda-ku, Tokyo 100–0001, Japan. For Akihito is none other than his excellency the emperor emeritus of Japan, although our running into him and his throng of armed bodyguards in the corridors of our museum dates from when he was still crown prince and his crab-studying father, Hirohito, was the emperor. But many community scientists of lesser stature face obstacles when they try to publish their results because lots of the established scientific journals require that authors have an institutional address.

POOLS OF WISDOM

"Publish or perish" is one of the better-known academic adages, and it applies as much to community science as it does to conventional academic science: you need to tell the world about your discoveries. Why? Because science, even if undertaken by monomaniacal loners, is very much a social enterprise. Information, ideas, and results need to flow unimpeded through the circulatory system of science so that everybody can, in Newton's words, see farther by "standing on the shoulders of giants." That is how science as a whole progresses. If you keep your findings to yourself, they are useless to science, and your discoveries may as well not have been made.

So scientists, professional and nonprofessional, need to document their discoveries, and the most common way to do so is still in the form of a concise scientific article or "paper." Although some papers are beautifully written, they are essentially standardized reports on a smaller or larger set of scientific discoveries. The typical "IMRAD" format starts with an Introduction, in which the background of the study is explained, followed by a section titled "Materials and Methods," where you describe exactly what you did and how, so that anybody can check whether you approached it the right way, and redo your study if they wish. Next comes a "Results" section, where findings are presented in a dry, neutral way without any pomp or exegesis. Then comes the "Discussion" section, where you finally get the chance to say what you think your results mean. There may be some peripheral sections, such as an abstract or a "Conclusions" section, and there may be appendices with lengthy tables filled with numbers, and, of course, a list of references to other publications, but in essence, this is what each and every scientific paper looks like.

In light of the uniformity of their content, it is surprising that there are so many different journals in which these papers are published: a whopping 30,000 of them worldwide. This is largely a legacy from the time that scientific publications appeared printed in journals, and each journal would serve a specialized, even hyperspecialized community of researchers. If you were a Russian biologist working on nematode worms, you would subscribe to the *Russian Journal of Nematology*. A limnologist interested in the freshwater fauna of the Pacific probably could not miss her latest issue of the *New Zealand Journal of Marine and Freshwater Research*, while a botanist working in West Africa would swear by the irregularly appearing *Études sur la Flore et la Végétation du Burkina Faso et des Pays Avoisinants*. Ecologists with a more general or theoretical interest would read *Ecology*, the *Journal of Ecology*, and *Plant Ecology*. Sometimes these journals would be published by academic societies, but often they would be in the hands of commercial publishers selling subscriptions to the growing global community of scientists, students, and university libraries.

And some publishers would abuse the publish-or-perish imperative to bleed science to death. Scientists needed access to the journals in their field to keep up to date. So they would take out a personal subscription or get their university library to buy one, often to the tune of thousands or, for the top journals that you really needed to keep tabs on, tens of thousands of dollars per year. But that is not the only cost.

A favorite anecdote among scientists is the response to telling their nonacademic friends or relatives that their article is going to be published in a scientific journal. "Wow, great!" their friends say. "How much are they paying you?" For unlike commercial magazines and newspapers, scientific journals (which have a much smaller print run) do not pay their authors—and frequently they charge them page charges. When your article is published, the journal sends you a bill for "publication charges," which can amount to thousands of dollars per article.

I say "when" your article was published, but I should say "if." Scientific journals rely on a peer review process. When you send a manuscript to a journal, asking whether the editor will accept it for publication, the editor will send it to two or more reviewers, experts in the field of science that your manuscript falls into. To allow reviewers to speak their mind, and to prevent fistfights at the next conference, reviews are often anonymous. Still, some fields of science are so small, and colleagues' hobbyhorses are so well known, that you can easily guess the identity of the reviewer based on the forthcoming assessment. Once an editor accidentally asked me to be a reviewer on a paper I had written myself—that is how small some fields are.

Based on what the reviewers have to say, your manuscript could be rejected (with some top journals this happens up to 95 percent of the time), or you could be asked to revise and improve it first. This whole process of peer review, editing, and publishing could easily take more than a year. But publishers never pay for all this work because the editing and reviewing is all done by scientists, usually for free, as a service to their scientific community.

This publishing system, in which the scientist (who is often paid from taxpayers' money) is charged thrice by commerce, has been on the way out since electronic publishing became possible and printed journals with their physical distribution were no longer an imperative. In the early 2000s, scientists, thoroughly fed up with the perversity of the system, began to cut out commercial publishers by launching large comprehensive scientific publishing platforms with public money. They are still called journals, and they still use the peer review system, but essentially they are simply gigantic pools of online scientific articles in which anybody can search and fish for what they need. And with anybody, I really mean anybody. This so-called "open-access" movement aims to make scientific results accessible for free to anybody—within academia and without.

The electronic journal *PeerJ*, for example, has, since its founding in 2012, helped nearly 20,000 articles from biology and medicine see the light of day. They are free to access, download, and be shared by anybody. The twenty-two people who work at *PeerJ* still need to get paid, so there is a one-off publication fee charged to the author, but that is it. No paywalls, no subscription fees.

In addition to these open-access journals, many other initiatives are making freely accessible the scientific literature that was published before the open-access revolution or that is still being published in the dwindling number of traditional commercially operating scientific journals. There is the Biodiversity Heritage Library, which digitizes classic natural history books and journals going back centuries. There is JStor, which does the same for more recent journals from all of the sciences. There is Google Scholar, which will find a pdf of any article if somebody, somewhere, put it online. There are even pirate websites like Sci-Hub, where you can find 88 million scientific papers, mostly from commercial publishers, that legally should be behind a paywall. To the young Kazakhstani neurobiologist Alexandra Elbakyan, the academic Robin Hood who single-handedly founded and manages Sci-Hub, free access to scientific information and the benefits it brings was more important than the restrictive laws of copyright. And then there is BioRxiv (pronounced "bio-archive"), where biologists place "preprints," drafts of articles that have not yet been peer-reviewed or published but that the authors would like to make accessible as soon as possible, either because they think other scientists need to be aware of the content without delay or because they want to solicit suggestions for improvement.

It is hard to overstate the difference that open access has made for the democratization of science. For example, when I worked in Malaysia in the early 2000s, the library in my poor, newly founded university in its quiet backwater of the country could afford only a handful of local journals, and we were cut off from most mainstream scientific literature. But once open access came along, my colleagues and I could finally have the same unimpeded access to the latest publications as everybody else. And during the COVID-19 pandemic we saw how, for better or for worse, anybody, not just paid medical professionals, could check out scientific publications firsthand.

MOUNTAINS OF DATA

In the wake of open access came the open data revolution. In the past, the raw data on which scientific studies were based were discarded or stored in desk drawers or university filing cabinets where nobody could access them, only to be thrown away decades later during a departmental cleanup. The open data revolution, which today most scientific institutions and community scientists subscribe to, dictates that those raw data (notes, measurements, images, readouts—all the strands of information on which scientific studies are based) have to be deposited digitally in a public and permanently accessible place.

Biodiversity data, for example—transcripts from labels in the collection of the Smithsonian Institution just as well as the moth sightings of Patje Debeuf—are deposited in open data banks such as the Global Biodiversity Information Facility (GBIF), where they can be retrieved by anybody, at any time, instantly and for free. Genetic data, those endless strings of A's, C's, G's, and T's that Sigrid Jakob is generating in her kitchen counter lab, are stored in GenBank. It takes only seconds to go to the GenBank home page and type "Sigrid" into the search box to quickly retrieve the long list of DNA sequences that she generated and deposited there.

I click on a random entry of hers and up comes a bit of string of about a thousand DNA letters from the mushroom *Entoloma carolinianum*, which she sampled on Staten Island, October 31, 2022. The record also mentions the link to iNaturalist where she placed a picture and the exact location where the specimen grew (on the shore of Orbach Lake in High Rock Park). If you go to that link, you can see that Jakob initially identified it as a member of a different mushroom genus, *Hygrophorus*, but another iNaturalist user, Stephen Russell (author of *The Essential Guide of Cultivating Mushrooms*), provided the correct identification, which was then adopted by Jakob and seconded by a third user, John Plischke of the Western Pennsylvania Mushroom Club. And who knows, one day somebody else may download and compile these and similar data to study, for example, how mushrooms colonize urban parks, or how the time of year that fruiting bodies appear shifts with climate change. All this flow of data and ideas would not be possible if scientific data were discarded, locked away, secreted behind paywalls, or disconnected and fragmented.

With open access to literature and data, community scientists in their self-built home labs or community maker spaces are as connected with other scientists, whether professional or lay, and with the greater world of science as are the paid scientists in their universities and research institutes. And while literature and data are the fuel that powers the scientific enterprise, its engine is asking the right questions and understanding the answers. And that is where MOOCs come in.

Massive open online courses took off in 2012 (dubbed "The Year of the MOOC") and have grown ever since. They are online courses in a broad or specialized subject, and they can be attended by anyone. There are video lectures, exercises, forums to interact with other students and lecturers, and all the other materials you would need to get a handle on whatever scientific field takes your fancy. On the MOOC platform Coursera, you just type in a search term—"microbiology," for example—and up comes a whole bunch of free courses on microorganisms offered by universities, museums, and companies from all over the world. If you type in the term "evolution," one of the MOOCs you will find is "Evolution Today," which two colleagues and I set up a few years ago and which now has been taken by some ten thousand people—many more than the combined number of students in all my university classes since I began teaching thirty years ago.

And while Coursera is a general-purpose site, you can find a collection of online courses more geared toward biodiversity and ecology enthusiasts on the Barcelona-based platform Transmitting Science, including courses in building and managing a natural history collection. And that is good news because, as we shall see in the next chapter, collecting specimens is an essential component of becoming an urban naturalist.

6

ON CHESIL BEACHCOMBING

On Chesil Beach is a short novel about the disastrous wedding night in the early 1960s of a couple on the romantic Chesil Beach on the UK's Dorset coast. The painfully beautiful story recounts how, in just a few hours, a concatenation of psychological obstacles caused by differing social norms, backgrounds, and expectations drives a permanent wedge between a young couple who are otherwise very much in love with one another.

When the book was published, in April 2007, the author, Ian McEwan, went on BBC radio to talk about the novella and his writing routines. He described to the interviewer how he had picked up three of the characteristic round, polished pebbles from Chesil Beach and kept these in front of him on his desk as objects of inspiration while he was writing his book. Hearing the interview, Dorset nature protection authorities pricked up their ears and, as one commentator wrote, reacted as if McEwan had confessed to loading one of Stonehenge's boulders into the trunk of his car and driving off with it. Chesil Beach, they exclaimed indignantly, is a Site of Special Scientific Interest, and no object may be removed from it—not even a handful of shingle, and not even if you're a Booker Prize–winning novelist. In the end, the scandal escalated to such a height that the writer was forced to return his inspirational rocks to the beach or be fined £2,000. He chose the former.

This poignant and rather sad story demonstrates the extent to which our relationship with natural history specimens has become distorted. True, McEwan is not a geologist, and he did not pick up those stones to label and identify them and keep them in any kind of scientific collection. But in essence he picked them up and put them in his pocket for the same reason that any of us, you and I, laypeople, community scientists, and professional scientists alike, preserve found natural objects: our imagination is piqued by their beauty, appearance, "realness," and the stories they tell or that we can tell inspired by them.

Though approaching sixty at the time, when McEwan took the pebbles he responded to the same impulse that makes a child take home a shell or a bone, a feather, or an insect: pure fascination with the endlessness of forms that the natural world offers us. By turning such beautiful, natural curiosity into a punishable offense, authorities are not protecting nature; instead, they are discouraging and turning away the very people who eventually may save it. This chapter is about the vital importance of building a collection—the seemingly old-fashioned but timeless enterprise of picking up, preserving, curating, and studying specimens from nature and using them like ever so many Rosetta Stones to reveal the secrets that nature holds.

BUG PERIOD

In the mid-1990s, besides being a postdoctoral researcher at a university, I worked as a freelance science reporter for a couple of newspapers and magazines. On one occasion, I had the good fortune to be sent on an assignment to interview the famous entomologist and ecologist Edward O. Wilson on a visit to Amsterdam. I met him in the lobby of his hotel, where he was slumped on a sofa. Slightly reluctantly, because he was tired from his flight and from the lecture he had just delivered, he allowed me to interview him. I could tell he was just going through the motions because he was mostly reciting sentences from his autobiography, *Naturalist*, which had just appeared, but I was happy with anything he gave me. One sentence in particular stuck with me: "When I was nine years old, I had what we call in America a bug period," he told me. "All children have a bug period—I just never grew out of mine."

I knew exactly what he meant. My own bug period similarly started when I was nine, but it had been preceded by a shell period, a fossil period, and a gemstone period, and was briefly followed by episodes during which my focus shifted to preserving pressed plants, dried mushrooms, and bird parts. I kept dead birds in our kitchen freezer, scaring the bejesus out of my mother whenever she opened it, to later cut off the wings and feet and dry them, and boil the heads in sodium carbonate to obtain clean white skulls. By the time I entered secondary school, all available shelf and wall space in my room was devoted to my nature study passions.

I had used thin iron wire to affix my dried, stretched-out bird wings to softboard panels that hung from the wall. In the spaces left between the huge wings of a great black-backed gull, a herring gull, and a shelduck I arranged the daintier wings of small waders and songbirds. (Unfortunately, they eventually all became infested with museum beetles, and my mother made me get rid of them.) In the corner of my room hung a plexiglass-fronted display cabinet that my father had built me, in which I kept all my snail shells, arranged by size and shape. Later, I realized I had mistakenly mixed land and sea snails, being misled by the general shell shape, which can be similar in unrelated families. And on a shelf, among my eclectic collection of nature and science books with dated titles like *Wanderings Near and Far*, *A Boy's Book of Radio Electronics*, *Seeing Is Knowing*, and *What's That Beetle?*, stood several ring binders with dried plants, and beneath them a growing pile of small fragments of leaves, flowers, and twigs shed by the brittle plants each time I took them out to show visitors around my "museum."

By no means was my room unique, though. True, mine was exceptionally cluttered, but whenever I would visit my school friends at home, they, too, would often have similar collections. Dead insects, shells, rocks, seeds, or bones were favorite hoardables for many children in those receptive years before their teens. But as they entered secondary school, one by one they would leave their bug period behind. Sports, music, and other more mature pursuits would push the desire to fondle old bones or bugs to the background, and teenage hormones would do the rest. Soon the boxes with shells and pinned moths would begin to gather dust, be moved out of the way to the back of the storage shelves, and eventually be thrown out.

But occasionally, as happened to Edward O. Wilson, somebody managed to avoid saying good-bye to their bug period and instead remained stuck in it forever. For me, I have to thank my biology teacher, Mr. Vestergaard, for that. A traditional naturalist but also a fierce unconventionalist, the compulsory, narrow school curriculum sat uncomfortably with him, and he would often get sidetracked during his classes and launch into long monologues about nature, ecology, conservation, birds, and his adventures collecting insects all over the world. Most of my classmates welcomed these because it meant they could just sit back and enjoy a lazy class—winking to one another and egging him on with the occasional "Then what happened?" or "Oh, wow!" and "Tell us more about that!" But I and a handful of other classmates were genuinely completely absorbed in the wonderfully exciting world of science and naturalist exploration that he opened up for us. Often, he would bring boxes with huge pinned tropical insect specimens to school, and we would pore over them and be enraptured by the horns of metallic rhinoceros beetles from Indonesia or the twenty-centimeter-long ovipositors of huge parasitic wasps from the Amazon. And (unbelievably, in hindsight), on the weekends, after a full week of teaching, he would still find the energy to take a small group of like-minded classmates on field trips.

On one of those trips, which were devoted to spotting birds, identifying plants, and any other nature-related subtopic we could think of, and punctuated by the tradition of his treating us to coffee and apple pie with whipped cream, I picked up a beetle. At the time, I knew very little about beetles, but I was certain I had never seen this one before. It was about two centimeters long with very thin, long, green-metallic legs, antennae, and underside, and on top a delicate pattern of off-white squiggly lines on a bronze background. It was a cool day in early spring, so the beetle was a bit sluggish, but it still tried (ineffectively) to bite into my fingers with its long, sickle-shaped jaws. I passed it to Mr. Vestergaard and asked him what it was. "Ah," he said, "that's a tiger beetle, *Cicindela hybrida*." And then a thought crossed his mind: "Would you like me to pin it for you?" My eyes must have lit up.

A week later, in the central hall of our school, he passed me a small wooden box with a cardboard lid. When I opened it, there was the same tiger beetle again, now in a dry and mounted state. From being a live, transient animal that had tried to bite my thumb, it had been transformed into a valuable, permanent, and equally beautiful scientific specimen. A long, thin pin had been run through its right-hand wing cover, with about one centimeter of the pin remaining above the animal so that it could be handled and moved around. A small label written with India ink in my teacher's hand had been affixed to the lower part of the pin stating the locality, date, and name of the collector (me!). Its legs and antennae did not point awkwardly in all directions, like the dead bugs that I kept in a box at home. Somehow Mr. Vestergaard had managed to arrange them symmetrically, with the forelegs pointing forward, the middle and hind legs pointing backward, and the antennae neatly curved over the head and onto the thorax.

Maybe I could catch and pin a few more beetles myself? Little did I know that that idle thought signaled my reentering the bug period for good. This single specimen of *Cicindela hybrida*, collected on an April day in 1978, became the nucleus around which I grew an ever-expanding collection. Although I could get tips and tricks from Mr. Vestergaard, in those days before YouTube tutorials and with a local library that understandably had a poorly stocked entomology section, I had to learn to be an insect collector the hard way.

Pick your poison, to begin with. Mr. Vestergaard carried in his field bag a thin-walled glass killing jar with a large rubber stopper and at the bottom a layer of plaster of Paris impregnated with cyanide that he derived from a secret source (we suspected the chemistry teacher). Unable to get access to cyanide, I had to find another poison. For a while I used jam jars with a few strips of toilet paper soaked in chloroform, which at the time I could still buy in the pharmacy (provided I satisfactorily answered that eternal pharmacist's question, "What do you need it for?"). Fortunately (for chloroform is now known to be carcinogenic), I was unhappy with the springiness and unmountability of insects killed with chloroform, so after a year I went looking for an alternative, which I finally got from another beetle collector, with whom I exchanged specimens by mail: ethyl acetate. A few drops on tissue paper in a soft plastic jar does wonders (don't use hard plastic; the ethyl acetate makes it melt): the insects die in a matter of seconds and the specimens stay fresh and pliable for a long time. Fun fact: ethyl acetate is the main ingredient of acetone-free nail polish remover, so I could easily and cheaply get it in any drugstore. Its odor was better than chloroform's, too.

It is, by the way, a common misconception that insects are killed by impaling them with the pin. The pins are used only as a way to handle the (previously killed) specimen without touching it, and for affixing labels. Insects' physiology is such that, unlike vertebrates, they would not die from having their body pierced—as I learned when one night I woke up because a large pinned chafer that apparently had not been in the killing jar long enough had come to and started lumbering noisily, zombie-like, pin and labels and all, through my bedroom.

Speaking of pins, this was another steep learning curve. Common household pins are not suitable for the job. They are too thick and too short, and they rust. What I needed were proper insect pins: some four centimeters long, made from stainless steel and available in a range of thicknesses. In the members' booklet of the Netherlands Entomological Society, I found the telephone number of the only entomological supplies company in the country, and, in exchange for a large chunk of my pocket money, gained access to an endless supply of insect pins. They came in sets of one hundred, nicely wrapped in a paper mini-envelope, confusingly stamped with a picture of an elephant and the words "Made in Austria."

But many beetles are only a few millimeters long, too small for even the thinnest (the near-invisible "type 000") of insect pins. These, I was told, first needed to be glue-mounted on a card, after which one end of the card could be insect-pinned. I could not afford premade mounting cards, so I made my own, spending long afternoons outlining six- by seventeen-millimeter sections (for the No. 7 standard mounting card size) on blank postcards and cutting them out to glue the smaller beetle specimens on. Glue is glue, I thought, so I took the solvent-based superglue that my father had on his desk. That this was yet another mistake, as I found out when the first question in one of the identification keys in my beetle book was, "Is the underside of the beetle smooth or roughly punctured?" My beetles had all been solidly and permanently affixed, belly down, to the mounting card, and there was no way to dislodge them for inspecting the presence or absence of punctures on their undersides. This is when I learned that small insects are mounted on boards with water-based glue only so that you can dissolve the glue and look at the bottom of the insect if needed.

In hindsight, despite the many trials and errors, learning the tricks of the trade of an insect collector was a lot of fun and extremely exciting to my young and eager mind. Throughout secondary school, my collection grew and grew, as did my entomological library and my knowledge of and passion for beetles. Within a few years, I was publishing scientific articles, knew by heart the Latin names and peculiarities of a few thousand beetle species, was identifying specimens for natural history museums all over Europe, and was corresponding by post with coleopterists in Germany (on both sides of the Berlin Wall—the love for beetles uniting us all), Denmark, Finland, Japan, and Russia. By the time I left school, my collection held over 15,000 specimens.

What exactly was the attraction? Undeniably, there were aesthetics, and all the friends and relatives to whom I proudly showed my collection agreed with this. My glass-topped insect drawers with their neat, regimented hierarchical arrangements of specimens into species, species into genera, and genera into families, all mounted in the same position and with minuscule handwritten labels, pleased the eye and the orderly mind. Also, the collection was like a scrapbook of mementos: as the author of *What's That Beetle?* wrote, "Every specimen reminds one of a pleasant day spent out in nature."

But unlike a collection of, say, postage stamps, a natural history collection is so much more. To me, it was what made me ask questions about ecology and evolution. How come there were only two species of water beetle in the pond in our garden but twelve in the ditch next to the soccer field? Why do the two dark gray fungus beetles *Apocatops nigrita* and *Fissocatops westi* differ in virtually nothing but the wildly different shapes of their penis? Why is there only one species of *Megasternum* (two millimeters long, spherical, shiny black, rummaging around in moist places) in Europe but no fewer than thirty-six species of the related and very similar-looking genus *Cercyon* (with the same size, shape, color, and habits as *Megasternum*)?

A collection of stamps or coins, cigar bands, or seventeenth-century clay pipes (all of which I admittedly also collected at some point in my life) does not make you ask such questions. Building a collection of such human-made objects is a goal in itself. There is no mystery there. If you want to know the origin of a certain stamp you can simply ask the post office. But each and every beetle specimen is a portal into science and can be used as a key to solving the mysteries of the natural world. My natural history collection, in essence, was what turned me into a naturalist. And you too can similarly reenter your bug period.

THE REAL DEAL

But before I guide you through the many fun ways in which you can build your own urban natural history collection, let me first get a few things out of the way. As you saw me raving in the previous section about the blessings of collecting, certain thoughts may have occurred to you. Thoughts such as "But isn't pinning bugs and pressing plants an outdated, old-fashioned type of biology?" Or "How can we be conservationists and at the same time kill the animals and plants we care about?" Perhaps also: "We can snap perfect pictures of everything we see, so why would we still want to put dead specimens in a box?"

Yes, one of the reasons people collected animals in the old days, before point-and-shoot cameras existed, was as a way of documenting their sightings. When the early nineteenth-century German naturalist Hermann Schlegel had a dispute with his mentor, Christian Ludwig Brehm, over whether a particular bird singing in the reeds in their hometown was a marsh warbler (*Acrocephalus palustris*) or just an inexperienced icterine warbler (*Hippolais icterina*), he simply shot the bird, took it to Brehm, and showed him that it was indeed the former. Brehm agreed that his student was right, slapped him on the back, and complimented him with the words, "You are a true naturalist!" (And pocketed the dead bird for his own collection.) A modern-day Schlegel would have taken out his smartphone instead, and a modern-day Brehm would have taken off his glasses, peered closely at the screen, using his fingers to zoom in on the image, and similarly have relinquished his untenable taxonomic position.

Indeed, birders today no longer shoot the objects of their study. In many parts of the world there are now almost more bird-watchers than birds, so the animals would quickly be driven to extinction if they did. But another main reason why collecting dead birds has gone out of fashion is the enormous increase in our ornithological knowledge. Thanks partly to all those eighteenth-, nineteenth-, and early twentieth-century bird collectors and their kilometers of shelves with stuffed birds, which are now in the world's natural history museums, we know so much about the various plumages of different birds, their distributions, and their migration routes. That information is available via hundreds of field guides with exquisite artwork, websites, and image recognition smartphone apps. And the birdwatchers themselves are armed with high-resolution digital cameras equipped with the best telephoto lenses. These days, bird identification has become, if not a breeze, then at least much easier than it was before. These days, therefore, ornithologists kill and preserve bird specimens only if they are sampling one of the few remaining remote areas where the avifauna is incompletely known, or if they suspect they have a new species on their hands. For the rest, observation and picture-taking do the job.

But that is birds. For most other groups of organisms, the situation is quite different. Take land slugs, for example. Just a slimy handful of the world's many thousands of species of slug can be identified based on external appearance alone. There is the Australian red triangle slug, *Triboniophorus graeffei*, with its helpful red triangle on its back. And there are perhaps a few more species that happen to have some unambiguous distinguishing marks on their body and can thus be identified from a photo. But almost all others are some shade on the gray-brown spectrum and, well, slug-shaped. They need to be preserved in alcohol, dissected, and have their internal organs examined before a slug expert is willing to give you an identification. The same is true for most other invertebrate animals, and even for many plants and some vertebrates, such as the tiny tropical so-called microhylid frogs (735 species and counting, most tiny and bumpy and bearing a mottled pattern in earth tones). At the moment (with the exception of most birds, mammals, and a few other small groups, together just a few tens of thousands of the world's many millions of species), almost the entire Earth's biodiversity is still so poorly studied that the only way to map species diversity and distribution is to collect, preserve, and investigate actual specimens.

This may change, of course. As we saw earlier, scientists are making rapid progress with the use of AI-based image recognition and DNA analysis to identify species without the need for a preserved specimen. Over the coming century, at least where the simple logging of the presence of a species at a certain place and a certain time is concerned, collecting may slowly become obsolete as it has already for birds. But there is still a long way to go, and that way is paved with curated specimens.

Besides, collections are important not just for mapping biodiversity. The date and location where a certain species was found are just the tip of an iceberg of important information contained within a specimen. Pressed plants in herbariums hold traces of the insects, fungi, and viruses that have attacked them before they were picked. Remnants of dung often adhere to pinned dung beetles so that it is possible to determine a species's preferred type of excrement. The gut content of fishes in alcohol can be studied so that we may know their diet. A wide and still largely untapped range of aspects of a species's biology can be accessed by employing natural history collections.

More important, collections let us look back into the past. Large, orderly natural history collections have existed since the early eighteenth century, so we can use them as biological archives to see how species and their ecology have changed over hundreds of years. In the natural history museum where I work, for example, I have used the land snail collection to show that the grove snail, *Cepaea nemoralis*, has evolved more brightly colored shells since the mid-twentieth century, probably as a result of climate change (the lighter shells reflect the sun's heat and protect the snail inside against overheating—we will come across this in a later chapter). In our herbarium collection, two students have studied insect chew marks on the dried leaves of the American bird cherry, *Prunus serotina* (imported into Europe from North America a few centuries ago). They discovered that, over 150 years, Dutch native insects have doubled their appetite for this toxic exotic plant (meaning the plant has become firmly incorporated into the Dutch ecosystems). The scientific literature is full of such collection-based research.

Okay, you might say, those are professional collections, amassed and maintained by large research institutions and national natural history museums. They are much more important than the puny, eclectic hobby collections of amateurs and private collectors. You might say this, but you would be wrong. The vast majority of the roughly 50 million specimens in my museum, for example, were collected by private collectors who, at some point during or after their lives, donated their collections. Our holdings are nothing more than an amalgamation of thousands of small and large private collections, mostly compiled by serious amateurs—people who in daily life hold other jobs but collect (and often become world-renowned specialists of) a particular type of organism. People like Piet Kanaar, an eye surgeon *and* world authority on clown beetles, or the housewife Christa Deeleman, a leading expert on cave spiders from the Balkans; nor do we forget Bram van de Beek, the Dutch reformed preacher but also owner of the largest bramble collection in the world. Most of the large natural history museums in the world have built up their collections in a similar way. That is the thing about any natural history collection, whether they be large or small, amateur or professional, local or exotic: every specimen, as long as it is properly labeled and curated, has permanent scientific value and will sooner or later be used by scientists to study biodiversity patterns over space and time.

Still, there is the issue of ethics. Does it really make sense to kill so many wild animals and plants for science in this day and age? Is that an evil that is really necessary? Well, it depends. First, it is important to underline that naturalists are not trophy hunters. The only thing they care about is the knowledge that can be gained by killing and preserving a specimen, and such an assessment always plays a role in the split-second decision about whether or not to collect a specimen. Whenever a fluttering butterfly is taken from the net, whenever a botanist's hand initiates a plucking motion toward a flower or a spider's life hangs in the balance over a vial with alcohol, the collector will make a snap decision as to whether the suffering and death of this particular organism are justified.

The outcome of that decision depends on a number of things. If the species is common and widespread, the risk of taking a few specimens for a collection is usually smaller than it is for rare and local species. If the species is a fast breeder, producing thousands of offspring, the individual will be easily replaced by new ones; if it is a slow breeder, the impact will be larger. In the case of many larger vertebrate animals, such as the birds I talked about, our knowledge about them is so vast, their populations are so small, and their breeding is so slow that killing is justified in only a few exceptional cases. But for insects and other invertebrate animals (which are the main type of animals that are still killed for collections on a large scale), this is almost never the case. Most of them have such vast populations and produce such massive numbers of offspring that sacrificing specimens for research makes no impact. In 2015, for example, a team of researchers studied whether the collecting of some three thousand wild bees every year in a nature reserve in Colorado had any impact on their populations. They found no effect. This becomes all the more understandable when you realize that the number of insects that a hundred insect collectors kill in a year is no more than what a single bat eats in a fortnight—not to mention the astronomical numbers killed by streetlights, or whenever somebody cuts the grass in a meadow, digs up a garden, takes the car for a spin, or plows a field.

But perhaps it is not the demographic impact of collecting that you worry about, not the ethics of pushing a species closer to a potential edge of extinction, but rather the ethics of killing a living animal per se. The closer our evolutionary relationship is with another animal, the more easily it can be included in our sphere of empathy: it is more likely that people will feel sorry for a mouse than for a mosquito. And as our culture becomes more refined, the empathy envelope is pushed wider and wider. A 2019 study confirmed that the fewer evolutionary branches separate us from an animal species, the more likely it is that we will include it into our thou-shalt-not-kill morality.

In the end, the breadth of that morality is an entirely personal choice, and I know several people who have resisted the desire to become an insect collector for exactly that reason. Still, even a personal zero-tolerance policy toward killing specimens need not be an obstacle to building a valuable natural history collection. Many objects can be collected without any creature being harmed. Think of seeds, empty mollusk shells, bones, feathers, exuviae (the shed skins of molting insects), hatched egg shells . . . or even entire animals that are already dead: roadkill or the piles of dead insects that accumulate in light fixtures, for example.

Whatever you end up collecting (and, after reading this chapter, what excuse do you have not to?), the benefits will always greatly outweigh the disadvantages. A collection is a great way to get a firm grip on at least a nicely cordoned off part of biodiversity. It will help you think about concepts like variability, species, ecology, and evolution. And it will help you discover your own environment in a completely new way.

And I have not even mentioned the sheer artisanal quality of preparing, mounting, and curating specimens—a subject I indulge in in the next chapter.

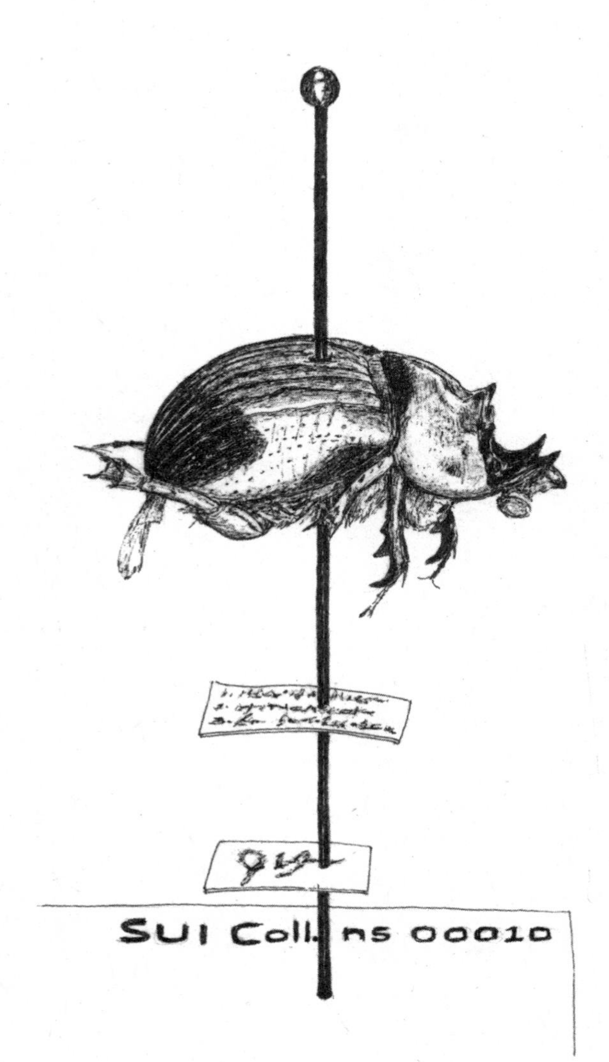
SUI Coll ns 00010

7

DEAD BUG BECOMES SPECIMEN

The insect mounting demo is always one of the most popular items on our community science expeditions in Amsterdam. I am seated behind a stereo microscope in a small theater in the east of the city while our group of community scientists crowds expectantly around me. It is midsummer break for the theater, so there are no shows, and we have been allowed to build our improvised lab there, a convenient staging area for our one-week expedition into the surrounding parks and other greenspaces. This morning we rode our bicycles to the centuries-old Jewish cemetery, a secluded, semiwild area separated by an ancient brick wall from the surrounding up-and-coming neighborhood with its smattering of modern office buildings. In the back of the graveyard, where the old, gray, partly toppled gravestones give way to wild brambles and willow bushes, we had placed some pitfall traps a few days earlier, and today we picked them up. That is why, in front of me, there is a white plastic tray with salty preservative liquid in which float, besides some leaves and twigs, a collection of beetles, bugs, centipedes, spiders, mites, and woodlice.

I explain what materials I have in front of me. First of all, the stereo microscope, also known as a dissecting microscope. It is a very different beast from the slide microscope that sits on another table. Unlike that one (which can magnify up to one thousand times and is used for looking at microbes or tissue samples), the stereo microscope has a much lower magnification (the type we are using on our expedition magnifies up to forty-five times), so it is suitable for examining small insects. Also, unlike the two eyepieces on the slide microscope, which are just for convenient viewing, the two eyepieces on the stereo microscope actually create a three-dimensional image. As your eyelids close around the soft rubber eyepieces it is as if you were, Alice-like, shrunk to the size of the bugs you have placed under them. It is almost, I say, as if you could touch the facets of the compound eyes, stroke the hairs on the back of the thorax, feel how sharp the spines on the legs are. My audience can see what I am seeing on the screen of my computer because the microscope image is fed not only to my eyes but also to a digital video camera attached to the microscope.

A good stereo microscope is indispensable if you wish to study small organisms like insects and other invertebrate animals, but also for the bryologist (who studies mosses), the lichenologist (who studies lichens), and those who investigate slime molds, plants with tiny flowers or spores, certain types of fungi, and so on. You can get a decent new one for the price of a midrange pair of bird-watching binoculars, but the really good brands often cost thousands of dollars. Fortunately, they tend to be durable, and you can cheaply buy second-hand ones of half a century old that are still in perfect shape.

Also in front of me are two pairs of forceps. Both are made from steel, but one is hard and sturdy, while the other one is made from thin, springy steel. The hard tweezers are fine for picking up robust specimens. I show this by lifting a large, two-centimeter-long bluebottle from the tray by one of its legs. But if I were to do the same with a two-millimeter aphid, I would crush it. With the soft forceps, however, I can pick up the tiniest insect without damaging it.

I place a small rove beetle in a small glass vial with some tap water to wash away the preservative. Then, I place it on a piece of tissue paper that I have pinned (with very short household pins placed at a sharp angle, so they don't get in the way) onto a piece of polystyrene. I pick up an insect pin as a tool to start preparing the specimen. From then on, everything I do, I do with my eyes fixed to the eyepieces of the microscope. It requires a lot of training, I explain to my audience, to gain good hand-eye coordination when you are looking at an object through the microscope while manipulating it with your hands out of sight. But with objects the size of this tiny rove beetle, there really is no other way: you won't see what you are doing with the naked eye.

First I turn the insect on its back. Then I very gently place the tip of my finger on one end of the elongated beetle. (Again, the amount of pressure you should apply is different for every type of insect and is something you have to learn by trial and error; be prepared to irreparably squish some specimens as a beginner—my audience chuckles.) With the tip of the insect pin in my other hand, I begin to fold out the tangled hind and middle legs, spreading them as far and as wide as possible. Sometimes I hook the tip of the pin under the tiny claws that most insects have at the end of their legs, and this helps pull the leg out. Then I turn the specimen around and do the same with the front legs, the antennae, and the palps (the little antenna-like food manipulators that crowd around the mouth). Finally, I press down the head and the thorax a bit further so that the whole body is nicely in one plane.

Knowing that this tiny rove beetle belongs to the genus *Atheta*, a group with a whopping one hundred species in the Netherlands that are all pretty much the same size, shape, and color, I know that I am going to need to carry out a dissection. Many insects that are very similar to one another and almost not identifiable on the outside can be easily recognized by the shape of their penis or the parts of the female reproductive system (in my book *Nature's Nether Regions*, you can read why this is so). From the shape of the abdomen, I can tell this one is a female, so I use the tip of the insect pin to create a little space between the upper and lower half of the last segment. Then I take a new, thin insect pin from the little paper envelope and, adopting a magician's air, tap the tip seven times against the steel frame of the microscope. My spectators look on with puzzled looks on their faces. I explain that by doing this, I have just turned the sharp steel tip into a minuscule hook.

I insert the hooked pin into the opening that I have just created in the abdomen, twist it ninety degrees around its axis, and then carefully pull out the female reproductive system. After three days in brine, most of it has turned into a yellowish mush, but the part I am looking for is the spermatheca, a tiny, hard but transparent organ shaped like a miniature balloon animal that the female uses for storing sperm that she may or may not use later for fertilizing her eggs. The shape of this organ is very different even in closely related *Atheta* species, so you can use it to get a reliable identification. This particular *Atheta* has a spermatheca that is a little inflated at one end and twisted into a tight spiral on the other end. I put the head of the pin into my mouth and then place the thus wetted pinhead on the dissected genitalia that are now lying on the tissue paper. The droplet of saliva makes the tissue supple so that I can easily separate the spermatheca from the rest of the tissue without running the risk that in the process it dries out and is launched out of sight by the springy tip of the insect pin. It would not be the first time something like that happens. As I explain to my wide-eyed audience, the walls of my lab at home are probably covered in all the microscopic insect genitalia that managed to get away.

Next comes the actual mounting. The beetle is still lying spread-eagled on its back on the tissue paper, its spermatheca next to it. I open a bag with small four-by-eleven-millimeter mounting cards. Nowadays I no longer cut them myself: we purchase them in bags of five hundred, made from stiff white cardboard with rounded corners and a few delicate gray-scale lines printed on one end. I use a toothpick to pick up a drop of water-soluble insect glue and spread a thin layer of this all over the card except the part with the scale lines where I hold it. Then I gently press the card, glue down, onto the beetle. When I lift it, the beetle comes with it, and I turn the card around and, again looking through the microscope, use the (unhooked) tip of an insect pin to spread all the limbs into the wet glue. The antennae, palps, and front legs go forward, the mid- and hind legs backward. And, of course, I position the body of the beetle in such a way that it is placed nicely in the middle of the card and parallel with the long edges. Some entomologists go so far as to create perfect symmetry on their mounting card (here's looking at you, Krefeld Entomological Society!), but I usually stop when all relevant parts are visible.

I place the card-with-beetle on a special wooden block with a hole in it, take a No. 4 insect pin, pierce the card where the scale lines are, and then push the pin down into the hole until it goes no further. Next I take from a different bag a transparent plastic mounting card of the same size and use a different toothpick to place a tiny bit of Euparal on one end, into which I drop the spermatheca. Euparal is a clear embedding medium that eventually will harden into a clear droplet. What I thus produce, essentially, is a miniature microscope slide, so that later, when I try to identify this specimen, I can use light coming from below to study the shape of the semitransparent spermatheca. This plastic mounting board is also pinned to the same pin, ending up a few millimeters below the card with the beetle on it.

Finally, I pick up one of the preprinted labels (I use acid-free paper for that). Peering closely at the miniature three-point font with which I have laser-printed four lines with the collecting details, I read out what it says: "The Netherlands, Province of Noord-Holland, Amsterdam, Jewish Cemetery, 52.360°N 4.946°E, in pitfall trap; 9 July 2021, Leg. Taxon Expeditions participants." I pin this label to the same pin and hold out the now finished specimen for all to see. Everybody peers closely at it. "What does 'Leg.' mean?" someone asks. I explain that this is short for *legit*, Latin for "he or she collected"; on a label, it usually precedes the name of the person who caught the specimen. In this case the whole group of citizen scientists is credited.

Later, when I have figured out which species of *Atheta* it is, I will affix a second label to the pin with the scientific name of the species and the person who identified it and when. For example, "*Atheta amicula*, M. Schilthuizen det. 2021" (where "det." stands for the Latin *determinavit*, "he or she identified"). From that moment on, this single insect pin has become a nugget of biodiversity information. It states that this species was found at that date and that location. The specimen itself is available for other scientists to study. And the specimen can be photographed, and the label information included into a biodiversity database. The international Global Biodiversity Information Facility, for example, holds such data for 2.2 billion specimens from natural history collections worldwide. And each of those data points has been brought into being by somebody, somewhere, carefully preparing a natural history specimen, just as I just did.

SPIDERS ON SPIRIT

The reason I have just spent five pages on a minute-by-minute account of the mounting of a single miniature rove beetle is to show that there is an artisanal quality to it. Over the course of tens of thousands of specimens, I have honed my (admittedly narrowly applicable) skills. I take pride in the specimens I produce and rest assured that they will be preserved for many years, if not centuries, to come, and will serve many generations of scientists that come after me, just as I am studying specimens that were collected by some like-minded predecessors many, maybe even hundreds, of years ago. And it does not really matter what kind of object you collect. Whether you build a collection of dry snail shells in clear plastic boxes, or spiders in vials with ethanol, or dried grasses fixed to herbarium sheets, you will similarly acquire a set of unique dexterities in preparing, cleaning, mounting, preserving, labeling, and organizing your specimens. And it is these skills that the participants in our community science courses usually enjoy so much: good, time-honored manual labor in the service of acquiring knowledge.

So there are many good reasons why building natural history collections should always be part of your community science projects. It produces long-term scientific specimens that eventually will become part of national or regional natural history museums and be used by scientists of the future. It also means that your work (the identifications of species, for example) can be verified by other people. And, possibly most important, it offers an opportunity to get really close and intimate with the objects of your study. You literally get to know them inside and out, and you begin to understand them at a different level than you would from observing them in the wild.

It is all the more surprising, then, that collecting seems to be going out of fashion. A mixture of reasons contributes to this state of affairs. Overzealous legislators, sentiments of empathy, fears of contributing to environmental damage, and the misplaced idea that collecting has been supplanted (rather than augmented) by more modern techniques all conspire to discourage people who otherwise might be inclined to preserve specimens. And museums, which, as we saw, largely rely on private collectors to keep their holdings up to date, are beginning to notice this. In 2022 the Academy of Natural Sciences in Philadelphia sounded the alarm about the sharp decline in specimens donated to it. And, in a similarly distressing publication from 2021, a team of entomologists analyzed the online databases of moths and butterflies in all U.S. collections and saw that from the late 1990s onward, the incoming stream of specimens had been drying up. This means that gaps are starting to appear in the biodiversity archives of voucher specimens that we have been keeping for centuries. In times of severe environmental upheaval, that is worrying, to say the least.

In the UK, such public nature icons as Sir David Attenborough and the BBC presenter Chris Packham are also voicing concern about the demise of the collector. In a 2012 interview for *Radio Times*, the two lamented the many obstacles that today seem to prevent what to them came naturally in their youths. "I'm out there all the time and I just don't see the boy that I was and you were," Packham said, to which Attenborough replied, "Yes, and . . . that is because it's no longer allowed, no longer legal, to be a collector. Now, . . . if you were to pick up a feather and put it in your pocket, it would probably not be legal. And not to be allowed to collect fossils . . . it's absurd."

Fortunately, at least in many parts of the United States, making insect collections is still an optional science project for high school students. And there are lots of cool online video tutorials that teach you how to go about putting together your own small scientific collection of insects or any other kind of organism. At least in terms of resources, today's budding collectors are much better off than I was half a century ago.

With the chapters in this first part of the book, I hope to have inspired you to become a collector, a community scientist, an explorer of your world, even (or especially) if your curiosity has not been dulled, your thirst for knowledge is as insatiable as Edgar Allen Poe's, and your playfulness like Mary Treat's. In the next part of the book, I hope to open your eyes to the possibilities that the urban environment offers the community scientist. Cities with their completely new, human-made environment offer entirely new natural histories, a new ecology, new behaviors, even new evolutionary processes that are every bit as exciting as "wild" nature, the traditional playground of the naturalist. Be that urban naturalist, and start harvesting the rich pickings that the urban landscape offers. In the following pages, I will show you how. And I will begin with what I call frontier habitats: unexpected urban environments that urban biologists have only just begun to understand and you will notice only once you have learned about them.

THE CITY IS YOUR GALÁPAGOS: THE URBAN AS THE NATURALIST'S GOLD MINE

We are acquainted with a mere pellicle of the globe on which we live. Most have not delved six feet beneath the surface, nor leaped as many above it. We know not where we are.

—Henry David Thoreau, *Walden* (1854)

8

HIDDEN RICHES

Eight men, official-looking, in suits, long coats, and Homburg hats, stand in a half circle in a back alley in Washington, D.C. They are peering down into a hole in the ground, about a meter in diameter, from which a ninth man is emerging. The looks on their faces are hard to make out in the grainy, black-and-white, century-old newspaper photo, taken around September 25, 1924. They seem to be doing their best to look stern and important for the photographer, but their faces betray amusement. After all, how often do you come across a hand-dug secret tunnel network a stone's throw away from Dupont Circle?

The extensive tunnels were discovered by accident, when they caved in under a truck that negotiated the narrow alley behind 1512 21st Street, Northwest. Speculation was rife, with explanations ranging from bootleggers to German spies, but it did not take long before an investigative reporter for the *Washington Post* found out who the real culprit was: Harrison G. Dyar, the previous owner of the premises behind which that truck sank into the ground. Yes, Dyar admitted when the reporter tracked him down; he had dug that tunnel. And also the one underneath his other property at 804 B Street, Southwest. For fun, and as a form of exercise. "Some men play golf," Dyar explained. "I dig tunnels."

It all began in 1906, when he was digging a flowerbed for his wife, he told the *Post*. By the time he reached several feet deep, he felt the uncontrollable urge to just keep going. In the end, his private tunnel network, parts of which had multiple, interconnected levels and were equipped with electric lighting and tiled walls, amounted to some four hundred meters and went down eleven meters below the surface.

Harrison Dyar was not a man of half measures. Not only did he hand-excavate a labyrinth underneath the nation's capital, he also managed, under a false name and unbeknownst to his Washington wife, to maintain a second wife and three children in the Blue Ridge Mountains. He founded, published, and edited several journals, including *Reality*, a magazine for the Bahá'í faith, of which he was an adherent. But all this, the digging and tunneling, the bigamy, and the religious and publishing exploits, were just side activities to his real passion and occupation, which was that of a naturalist. Studying moths and butterflies, mosquitoes, and sawflies, he published 650 scientific papers, discovered and named some 3,650 insect species, and donated 44,000 insect specimens to the Smithsonian Institution, where they still are. His 1890 three-page article with a method for determining the number of molts of caterpillars is still being cited every week somewhere in the scientific literature. As Mark Epstein, who had the self-appointed and enviable task of writing the biography of this man for whom even the word eccentric is too bland a label, writes, "Everything he did was in the hundreds or thousands."

A large chunk of Dyar's plethora of pinned insects can be viewed on the website Bionomia. As we saw in a previous chapter, more and more scientific data are being made publicly accessible in the interest of open science, and Bionomia is one of them. To give credit where it is due, it links natural history objects from all over the world to the naturalists who collected and identified them, and Harrison Dyar is also there: for a few thousand of his specimens, mostly mosquitoes, you can look up the label details. And a few of those specimens have latitude and longitude coordinates that correspond closely to where his tunnel networks must have been. I can just imagine *Culex pipiens* mosquitoes breeding in a forgotten, water-filled cement pail in one of Dyar's shafts, and being encountered and collected there by him during a bout of excavation.

Even though Dyar may have opportunistically collected a couple of specimens in his tunnels, other urban naturalists purposefully enter the city's underbelly with the explicit aim to explore it biologically. Not by digging down themselves but by entering the subterranean world that already exists there.

Subterranean biology has always been a super-romantic, adventurous branch of nature exploration. Descending into that forbidding, dark, damp world where animals have been evolving in isolation for millions of years in an environment devoid of light and seasonality, and sustained on very little food—the possibility of finding completely unknown life forms in underground chasms where no human has set foot before, the sheer Indiana Jonesness of it, has always charmed naturalists, myself not excepted.

I can trace my fascination with caves and cave life to a children's book by Godfried Bomans, about a group of kids on a school trip to the Pyrenees who got lost in a cave and had to survive on candle grease. Aged six or seven years old, I devoured this book one rainy Wednesday afternoon at our dinner table (I must have borrowed it from the school library). The thrill was not just the expertly told story (a scene in which the children come across the skeleton of a missing speleologist, his characteristic red cap still stuck on the skull, is etched in my mind), nor was it that the book was the kind of forbidden fruit intended for schoolchildren twice my age. No, the main reason why the story made such a deep impression on me was that it fired my own imagination, which began to run wild with dreams of one day exploring such places myself.

Despite my living in the swampy Netherlands, where caves are as rare as cavities in hen's teeth, those dreams did come true. Ten years later, at the even more impressionable age of sixteen, I joined a group of hippy biology students on a caving weekend in the Ardennes (in a rickety Lada Niva, with The Doors on the cassette player the whole time) to learn the basics of speleology, and I have been dabbling in spelunking on and off ever since. I say dabbling, because my wife Iva is the true cave biologist of the family: she has gone down into almost four hundred caves, calls herself Cavernella on the internet, is an official National Geographic Explorer, knows how to scoot up and down sinkholes on a rope (I don't; I go on all fours and stop when it gets too steep), and did her doctoral studies on the evolution of the Anthroherponina, bizarre cave beetles from the Balkans.

Dabbling though it may have been, cave exploration has always been a part of my naturalist life. Together, Iva and I have investigated cave animals in Albania, Bhutan, Borneo, Bosnia, Georgia, Greece, Italy, Japan, Montenegro, Serbia, and Spain. When I lived and worked in Malaysia, I tracked down and explored caves in the forests of the interior that no biologist had ever entered before. Willfully breaking the iron law of caving never to enter a cave alone, I spent days charting caves by myself, squeezing through tight passages, swimming across underground lakes, discovering new species of cave animals, sampling their DNA, mapping out their populations, and making sketches and notes in a waterproof booklet I kept in the back pocket of my jeans. Being there by myself, deep underground, enveloped in a silence that was only punctuated by the drip of water from the ceiling—it was pure bliss.

But it was also pure science. Caves and other cavities that are formed naturally in limestone by percolating groundwater (a landscape referred to as karst) are extreme environments. As one moves from a sunny, lush forest into the dark cave interior, almost everything that is important in the life of a wild organism changes drastically. The light intensity decreases, the temperature stabilizes, humidity goes up, plant life disappears, any kind of food becomes scarcer. Within those one hundred meters or so between bright daylight and total darkness is compressed a multitude of narrow life zones, each with a specific set of ecological characteristics and organisms that can live there.

Since spelunking took off in the nineteenth century, speleobiologists have distinguished three broad categories of animals that you can find in caves. Trogloxenes (in Greek, *trogle* means "hole," and *xenos* "stranger," so "cave strangers") are animals that may enter caves but need to spend at least part of their life aboveground to complete their life cycles—you find those mainly in the entrance. Troglophiles (*philos* = "friend") are those whose natural niches are shady, secluded places, but they are not adapted to life underground to the extreme: they usually still have eyes and pigmentation. Troglobites (*bios* = "life"), finally, are the true cave dwellers of lore: blind, pale, wingless, often with elongated legs and antennae to "see" across a distance and make up for the lack of eyes, and many other outlandish adaptations in their shape and physiology. In the twilight zone between the outside and the cave interior, you may encounter all three categories, but as you go deeper, the proportion of troglobites increases at the expense of the other two. (This classification applies to land-based cave animals, but it goes for freshwater cave dwellers just as well—except that those are called stygoxenes, stygophiles, and stygobites, respectively, after Styx, the mythical river of the Underworld.)

It is not for nothing that caves have been called "simple natural laboratories." Since conditions for life there are so stringent, and since the ecosystem runs on what little food percolates in from the outside, caves harbor only a few species, and together those species form food chains reduced to their bare essence. In many caves, even after days of exploration, you will not find more than ten or twenty species of animals. So figuring out how the ecosystem works (who eats whom, who competes with whom) is much easier than in the outside world, where biodiversity is so much greater and the web of life so much more tangled and complex. Another fascinating aspect of cave life is that it is confined within the network of caves, crevices, and cracks that the weathering of the bedrock has created. Sometimes those confines are very tight, and a cave ecosystem is limited to a single hillside, and all the species that have evolved there have always been restricted ("endemic") to that one place. In other cases, an entire region is riddled by underground passages, streams, and a mesh of cracks too small for humans (but not for small cave animals) to enter, and you will find the same ecosystem wherever you enter a cave in that area. In other words, cave biology has all the hallmarks of traditional natural history, conducted in remote, wild natural places.

KARST OF CONCRETE

Remote and wild—so, never did Iva and I expect we would don our caving suits and headlamps right in the city where we live. And yet here we find ourselves in the medieval city center of Leiden, about to enter a cavern we had no idea existed, with archaeologist Jasper van Kouwen. He has been documenting and studying the underground spaces hidden from sight beneath the buildings of the old center of our city, and with his friendly demeanor he has built a network of house owners, concierges, janitors, and storekeepers who are only too happy to give him access to the underbellies of their properties. As we follow him through alleyways and under vaults and bridges, he enters houses here and there, cheerfully waves hello to the owner, and heads straight for an old wooden door or an archway, portals to a subterranean world most city dwellers do not know exists.

Among the places where we find ourselves is one of the oldest spots of the city, the open space south of the gothic Pieterskerk, where the Pilgrim fathers lived in the early seventeenth century and where five hundred years before that the count of Holland built a prison. At that time, Leiden was little more than a row of wooden houses built on river dunes along the Rhine, a few hundred meters away, and the brick prison tower lay in a swampy hinterland. Today that twelfth-century tower has become embedded in a complex of buildings from subsequent periods, and as recently as the mid-nineteenth century public hangings and canings were still conducted on the square in front of it. Kouwen says hi to the concierge (the building is now owned by the university's law faculty, and the only prisoners are the students that we spot in the library, captivated by their textbooks) and takes us down a staircase and through a heavily bolted wooden door. We descend into a dark dungeon where, a few meters below ground level, the erstwhile swampy environment betrays itself in the form of stalactite-like encrustations born of the chemical reactions among brick, mortar, and the groundwater that seeps through the walls.

Various species of spider have made cobwebs in corners of the vaulted ceilings of the old prison cells among engravings of people once incarcerated here (we spot some French Huguenot surnames, etched in quaint capital-letter orthography: MABILLE, DE LA CROIX, LARRUY), and remains of darkling beetles (*Blaps mucronata*, the churchyard beetle) litter the dust around the crude brick washbasin where the prisoners would conduct their ablutions. To Iva and me, the whole experience—the crumbling damp walls, the sound of our footsteps echoing in the chambers, the still water in deeper parts of the dungeon, and the troglophilic spiders and insects that we find while crawling through narrow spaces as we illuminate them with our headlamps—is surprisingly similar to the explorations of real caves that we have done far away from the urban world.

Like Leiden, many cities have a similarly ancient core that is permeated with such underground human-made spaces dating back centuries or even millennia. Lacking such relatively modern inventions as piped water, a closed sewage system, or refrigeration, the urbanites of old used the underground for those facilities. They dug wells to access and draw up groundwater, cellars for keeping food and drink cool, cesspits for waste disposal, and cisterns for storing rainwater. As cities grew, open streams were covered over, old buildings were torn down and new ones erected, and underground spaces were covered up and forgotten. The Italian city of Catania, for example, the largest city on the island of Sicily, founded nearly three thousand years ago, has been employing an urban speleology team for the gargantuan task of charting all the underground spaces in the center (not least because they pose the danger of collapsing in this earthquake-prone region). The map that resulted from this effort is not something you should view if you suffer from trypophobia, as it looks as if somebody traced the modern street plan with a hole puncher: under the surface there is a parallel city of pits, holes, cellars, tunnels, and passageways, interconnected and constructed side by side and one on top of another.

In many cities, the underground is even more perforated as the bedrock on which the city sits was used to provide the material that the buildings aboveground were constructed from. The soft limestone underneath Odesa in Ukraine, for example, was tunneled by centuries of Odesans to excavate the blocks of stone for assembling the city above. Effectively, Odesa sits on top of a negative image of itself in the subsurface: a network of more than 2,500 kilometers of tunnels—snaking, looping, and anastomosing—from which chalk blocks have been hewn to build the houses that line the streets above. Although underground quarrying still continues in a few parts, most of this unmapped labyrinth (which dwarfs the labyrinths of Paris and Maastricht by tenfold, not to mention the famous catacombs of Rome with their puny fifteen kilometers) lies empty. In modern times, it has been appropriated for everything ranging from accommodating illegal sewage discharge, mushroom nurseries, and the dumping of murder victims to serving as smugglers' hideouts and wartime air-raid shelters.

Like the catacombs of Paris, those of Odesa are also a popular playground for urban speleologists, or "cataphiles," a subculture of urban extreme sports, who explore and map the sheer endless passages, many of which are flooded and officially out of bounds. In Paris (or rather underneath Paris) the cataphiles are actively pursued by a special branch of the police (nicknamed "cataflics") that was founded in 1955 to keep the underground spaces safe and purged of illegal activities, but in Odesa it is a free-for-all. And that includes subterranean naturalists as well.

One of those is Oleg Kovtun. For several years, Kovtun has been scouring the catacombs for animals. Bats turned out to use them as roosts in their thousands; he found six species of troglophilic spiders; and he discovered a snail, *Oxychilus translucidus*, that had not been found in Odesa before. In 2014, while passing through a flooded tunnel underneath the intersection of Poshtova Street and Rozkydailivska Street, he dipped his water net into the murky waters and made his greatest discovery yet: a nearly blind, pale, one-centimeter-long freshwater shrimp that, as he later found out, belonged to a completely unknown species. In a publication in the journal *Subterranean Biology*, he and his colleague Dmitry Sidorov named it *Synurella odessana*.

But wait a second: the catacombs of Odesa are just a few centuries old! As we shall see in a later chapter, urban animals can evolve and adapt very fast, but surely a regular shrimp cannot evolve into a stygobite (without pigmentation and its eyes nearly gone) in such a short period of time? Indeed, as Kovtun and Sidorov point out, *Synurella odessana* probably infiltrated from a much older subterranean realm: in many places, the man-made catacomb system connects with natural caves and cavities in the limestone rocks, and the urban ecosystem in the catacombs is partly fed by the cave animals that have been living underground for millions of years and whose area humans have kindly extended by offering new spaces under the city for them to colonize.

The same is probably true for another troglobite that was discovered smack in the center of India's most populated city, Mumbai. Back in 2005, Tejas Thackeray, who runs the Thackeray Wildlife Foundation in India, heard a tale about "pink worms" that were living deep inside water wells in Mumbai. Not completely willing to dismiss this rumor as a fable, Thackeray would occasionally check water wells whenever he and his team ran across them, but nothing resembling a pink worm ever turned up.

Nothing, that is, until December 2019, when he peered into a twelve-meter-deep well in the courtyard of a school in the Jogeshwari West district and spotted a reddish, eel-like fish at the bottom. After a two-day effort, he managed to capture a specimen and immediately got in touch with his ichthyological colleague, Praveenraj Jayasimhan. Altogether, they were able to net five of the strange-looking eels: around twenty-five centimeters long and only half a centimeter wide, bright pink, completely devoid of eyes but with a very well-developed set of sensory pores along the head (which fish, and particularly blind fish, use to register pressure and vibration in the water). In other words, a stygobite, a new species that Praveenraj, Thackeray, and their colleagues published under the name *Rakthamichthys mumba,* after the city in which it was discovered. Apparently, says Praveenraj, the species leads an unobtrusive life in the city's groundwater subterranean ecosystem. Fittingly, the school where the blind fish was discovered was a school for blind children.

Even the underground spaces in Leiden that Van Kouwen guided us through have their own stygobites. The whole coastal part of the Netherlands is an estuary. The nearest bedrock is hundreds of meters below the surface and covered by layers upon layers of peat, clay, and sand; no natural caves are to be found anywhere. Yet blind freshwater shrimp somewhat related to the ones found in the Odesa catacombs live in the groundwater under Leiden, too. The enigmatic subterranean *Niphargus aquilex* has been pumped up twice in old Leiden water wells, once in 1852 and again in 1983, just a stone's throw from where I live. I am still on the lookout for that species.

Real karst in natural environments is not just limited to the large spaces that humans can enter and call caves but also includes much smaller fissures, down to capillaries in the rock so narrow that only a mite or springtail can move through. The same is true for urban karst. Urban biospeleologist Al Greene recently retired from the Public Buildings Service in Washington, D.C. During his many years as entomologist and pest control expert he spent his career looking for insects in any underground urban space that he could get his forceps into, and he has developed a richer picture of what constitutes the urban karst landscape than anybody else.

Sure, he and his colleague Nancy Breisch write in an article in *American Entomologist,* there are all these obvious underground spaces, such as "parking garages, subways, basements and machine rooms, steam and other utility tunnels, electrical vaults. . . ." But there is so much more because successive generations of contractors placed water pipes, power cables, fiber-optic cables, and air-conditioning conduits to connect all these spaces. Referring to an indeed horrifying picture of the spaghetti-like tangle of old pipes and cables (made from or wrapped in PVC, iron, rubber, asbestos, lead, and cloth) at a construction site in Manhattan, he points out how "short-sighted design, poor-quality construction, nonexistent maintenance, and uncoordinated jurisdictional authority" plus "poorly sealed utility penetrations and masonry joints" create many more underground spaces in cities than we would think. On top of this, there is the natural decay of concrete, especially around corroding reinforcing steel rebar, which generates an unmappable, dynamic wilderness of continuously growing fractures that penetrate the concrete matrix. The essential humidity is there too, thanks to the "discharge from leaking pipes or seepage through soil and fractured concrete."

Jointly, all this creates an urban karst network of spaces of all sizes that is, to all intents and purposes, of the same complexity as that found in natural karst rocks. "Deep infrastructure" is what Greene calls it and he is convinced that it connects nearly any underground space in the city with all the other ones. And it is these connections that urban subterranean animals move through. Animals like the American cockroach, *Periplaneta americana*, for example (despite its name, its original home is Africa and the Middle East), lives in cities all over the world—if the city is in too cold a biome, they stick to artificially heated parts of the urban karst. These troglophiles can have huge populations in sewers, living on the abundance of organic waste that accumulates there, but from there their populations send out tendrils through the deep infrastructure into basements and kitchens, and through pipe chases and elevator shafts manage to colonize homes on the top floors of skyscrapers.

Another unwelcome denizen of the urban karst is *Aedes aegypti*, the mosquito that transmits dengue, yellow fever, and zika. At the time mosquito expert Harrison Dyar was gophering through Washington's subsoil he would not yet have encountered it, but in 2011, after decades of climate warming, this tropical species was found breeding in the urban karst under Capitol Hill, where its larvae are protected against cold winter temperatures, and the adults happily emerge from drain covers to bite people at street level.

Speaking of street level, that is exactly where we find another urban extreme environment. And it is one that we are even less aware of than we are of the urban karst.

THE GROUND WE WALK ON

The richest biodiversity in cities is not always found where you would expect it. You might think that in the center of Paris, one of the larger city parks would claim the honor—perhaps the Jardin des Plantes next to the natural history museum, or the Parc des Buttes-Chaumont with its lake, cliffs, meandering streams, waterfall, and thickets of a variety of deciduous trees full of nesting birds. But in fact the most biodiverse spot in the city, as far as is known, is in the south, in a little-visited corner of the 15th arrondissement, wedged in between the train tracks that lead toward Orléans from Montparnasse station, the busy Boulevard Lefebvre, and the second-hand book market stalls at the former Vaugirard slaughterhouse.

There you will find a hidden gem: a wild ecosystem with well over a thousand different recorded organisms. It is truly a jungle, with an abundant tangle of vinelike *Melosira* with that nice giraffe-like green pattern on its stems, and scaly Cercozoa feeding on them. Underneath there are thickets of golden branched *Dinobryon* and the delicate flowerlike *Synura*. Everywhere you see beautiful reddish orange *Serratia* and *Dietzia*, and the fluffy dark yellow *Gordonia*. If you penetrate deeper, you will find the black velvety cushions of *Knufia* and *Bradymyces* fungi, and here and there a *Hartmannella vermiformis*, chasing prey with its thin tentacles.

Let me tell you how you can find this miniature urban rainforest. From the metro station Porte de Vanves you walk west and turn right, and then at the Afrikmarket turn right again onto Rue Castagnary. Keep on the right-hand side of the street, and walk down for about seventy meters. On your right you will see a modern apartment building with orange glass balconies and downstairs a patterned concrete wall and doors covered in wooden planks. There might be two small green plastic Dumpsters outside and the taxi of Mohamed Mekacher. Take two steps away from the building until you reach the granite curb and the rusty iron drain cover. There, squat down and peer at the cobblestones in front of the curb. See that slimy greenish-blackish film covering the stones? That is what is called a microbial mat, and it is where biologist Vincent Hervé found 1,169 species in June 2015.

Of course, those species are not birds and bees and flowering plants. The organisms I listed above are minuscule diatoms, protozoa, and bacteria, and you would need a microscope to see them. In fact, Hervé never even saw them. He simply sat down, took out a toothbrush, and scrubbed some of the green gunk off the cobblestones. Back in his lab at the University of Neuchâtel, Switzerland, he extracted DNA from the sample, conducted PCR of DNA barcodes, and sequenced them, as we discussed earlier in the book. Some serious number crunching and automated comparison with the reference barcodes in GenBank then led to a list of 1,169 species of diatoms, fungi, and a whole menagerie of other single-celled creatures. In fact, he collected not just from Rue Castagnary but also from eighty-nine other gutters all over Paris, wherever he saw a layer of greenish-reddish-blackish stuff, usually near a drain, gracing the surface of the street.

To the uninitiated pedestrian, these slimy layers of "dirt" are just a familiar feature of soiled streets that they would either be indifferent to or slightly repulsed by—something to be cleaned by the municipal vehicles that scrub the streets, but in any case not something one would associate with urban nature and biodiversity. But to the microbiologist, these microbial mats (one of the hottest topics in microbiology) are unicellular rainforests. They are maintained by the sun, which provides energy for photosynthesis and by water—from precipitation but, in the case of Rue Castagnary, mostly from the gutter-bound torrents that the municipal authorities unleash each morning and that are such a familiar sight to early commuters or Parisians returning home from a party.

These few millimeters of vertical space provide a mosaic of chemical and physical subtleties so rich and diverse that hundreds, even thousands of species can find a niche. The three-dimensional structure is like that of a rainforest, too. There is a top layer of green diatoms and blue-green algae that absorb most of the sunlight and that house bacteria that need oxygen. Other layers deeper down are where photosynthetic microorganisms live that capture the remnants of sunlight but that also house bacteria and fungi that thrive on waste products trickling down from the layers above them. Many of the bacteria exude a slime that, along with the filaments, gives the microbial mat its strength. Unicellular predators elbow their way through these thickets, prowling for bacteria and other smaller microorganisms to feast on.

There is also a day-night rhythm. Some blue-green algae need low light conditions and cannot stand strong UV light, so in the morning they migrate down to deeper layers and in the late afternoon, when the sun has set behind the zinc rooftops of Rue Castagnary, they crawl back up, akin to what some monkeys and other arboreal mammals do in a real rainforest, swinging high up in the canopy in the morning, relaxing lower down in the afternoon. And in the ebb and flow of signaling molecules that all these microbial mat organisms send out to each other there is even a chemical equivalent to the insect-frog-bird chorus that is such a typical acoustic backdrop in rainforests.

Although biologists (or at least the microbe-studying kind) have been fascinated by microbial mats for several decades now, their in-depth study has received a serious boost only recently, when it became possible to use machines that can read millions of DNA barcodes in one go to figure out exactly which and how many organisms cohabit in them. There is also a practical reason for this interest: biofouling.

Think of the historical center of an old city—Rio de Janeiro, for example. There are splendid historical churches from a bygone era, such as Igreja da Candelária, with its façade of local augen gneiss. When it was built, back in the seventeenth century, the front of the church must have looked a pretty light pinkish gray. But today it is a drab leaden color. No wonder, you might think: generations of urban street dust whipped up from the fourteen-lane Avenida Presidente Vargas, mixed with the exhaust fumes of the millions of combustion engines with the added-in pollution from the industrial city of São Gonçalo across the bay, would have covered the masonry in a thick patina of grime.

But when you start carefully peeling away these darkened layers and putting samples under the microscope, you will notice that it is not just a layer of dust particles. Instead, it is yet another microbial film, composed of a specialized ecosystem of algae, fungi, and bacteria that have made this rocky, sun-exposed and windswept surface their home. Photosynthetic blue-green algae penetrate deep into the rock, following crevices dissolved by the acidic waste products of rock-inhabiting fungi and bacteria. Sure, there is dust there as well, captured by the fungal mycelium and the slime produced by bacteria, but most of the discoloration you see is actually the organisms themselves: many of them produce yellow, orange, brown, gray, and black pigments. On the other hand, there is also an interaction between the façade-surface ecosystem and air pollution: lead and sulfur from exhaust fumes, for example, promote the growth of sulfur-eating and heavy-metal-tolerant bacteria.

To really come to grips with the communities of microorganisms that the outside of Igreja da Candelária is blessed with, a team of urban microbiologists from Universidade Federal Fluminense in Rio and the University of Oklahoma in Norman scraped bits of crust from the church's façade, and subjected these to electron microscopy, chemical analysis, and DNA barcoding. The numbers of species of fungi and bacteria they found put those from the Parisian gutters to shame: over 100,000 different species live shoulder to shoulder in the centuries-old microbial mats that clothe the gneiss church surfaces.

A notable proportion of those, say the researchers, are inhabitants of extremely hot, dry, and salty substrates—in other words, the microbial mat is not unlike those you would normally find in the crusts around a hot spring. That is not surprising, given the tropical location of Rio and the exposure to the sun and the salt spray from the bay. To protect themselves against UV radiation, these microorganisms are also rich in melanin and other dark pigments, giving the church its gloomy appearance. And then there are the gypsum and salt crystals that showed up in the electron microscope images. These had formed in the rock's top layers: minerals that precipitate out of the sulfur and calcium in the polluted air and the sodium and chloride from the sea spray. As these crystals grow, they force apart the upper layers of the gneiss, creating new spaces for the microbial mat to grow into. Moreover, the fungi and bacteria themselves also produce acids and other compounds that speed up the weathering of the rocks. In other words, the church's skin is eating itself up, thanks to the rich biodiversity of God's smallest creatures, which slowly but surely reduce the tough gneiss to ashes and dust.

What we have discovered in this chapter is that the urban environment is full of nonobvious but very novel environments that are just as worthy of exploration as their wild counterparts. And, because these extreme urban environments are so new and poorly explored, they are also places where new life forms can be discovered. In the next chapter, I show how the city is just as suitable a candidate for the discovery of new species of wildlife as are remote natural wildernesses.

1.0 mm

9

NOV. SPEC.

For more than fifteen years after I graduated, my work environment was strictly academia—until, that is, 2006, when I joined the natural history museum Naturalis, in Leiden, the Netherlands, where I still work part-time. Even though the kind of work I do at the museum—research, publishing, and teaching—is very similar to what I did in my university jobs, there is one very important difference: a live audience. Unlike the ivory tower of the university, in the museum I am constantly exposed to a general public that apparently likes the sort of thing my colleagues and I are doing. Every time I enter or leave the museum, whenever I briefly pop out for lunch, and even when I drop in to pick up a letter or hand some specimens to a colleague, I have to wade through throngs of visitors, who gawk at the skeletons of dinosaurs and mammoths, pore over the display cases with pinned insects, or are mesmerized by the video presentations of voyages of discovery in the days of old.

Such casual encounters aside, there is one dedicated place in the museum where the researchers and the public come face to face: our LiveScience hall. In this unique exhibit, which faces the street and can be entered without buying a museum ticket, visitors can witness live dissections by my anatomy-savvy colleagues Becky Desjardins and Liselotte Rambonnet, or provide hands-on assistance to malacologist Anthonie van Peursen, who always has a stack of clam shells in front of him that need sorting. Upstairs, the constant and slightly unnerving whine of a dentist's drill and the booming voice of paleontologist Martijn Guliker betray the live preparations of fossils from the museum's dinosaur dig in Montana, and every day there are several informal talks by the museum's curators and scientists about their ongoing research.

Some of the most frequently asked questions after such a lecture have to do with discovering new species, which is the bread and butter of many museum scientists. "How do you discover a new species?" is one of the common questions from the audience. "How do you know for sure it is new? Is there some international supervisory committee that you have to send it to?" Clearly, the discovery of something utterly and entirely new to science, an animal, plant, or fungus that had been overlooked during centuries of exploration, speaks strongly to the imagination.

Discovering new species of wildlife is something that sounds like peak exploration and conjures up images of adventurous naturalists scaling tall mountains and penetrating deep into remote rainforests to reach lost worlds where no scientist had ever set foot before. Yes, you will find new species in such places. But you can also discover new species—species not yet described and classified by scientists, that is—on your doorstep. As we already saw when we met Sigrid Jakob, who investigates mushroom DNA around New York City, it is mind-numbingly easy to discover undescribed species of fungi even in the heart of a metropolis. Science has discovered and described some 150,000 species of fungi, but even the most conservative estimations predict that there are at least ten times as many in existence. The same is true for insects, mites, nematode worms, spiders, snails—everything except perhaps those overstudied vertebrate animals. So even in the most visited urban park of the capital city of the best explored country in the world, it is a piece of cake to come across a hitherto undescribed species.

That is why my organization, Taxon Expeditions (a "scientific travel agency" that organizes real scientific expeditions all over the world for everybody to join), on an urban expedition in Amsterdam in 2019 dared to guarantee that we would be discovering new species. The expedition took place in the Vondelpark in the heart of Amsterdam and the explorers were a mixed team of interested people from the neighborhood on the one hand, and international biodiversity specialists on the other. We spent one week setting insect traps in a secluded corner of the park and studying the specimens in a makeshift lab that we put up nearby. Within two days, the wasp specialist on the team, Kees van Achterberg, announced that he had discovered a new species: a three-millimeter-long blackish parasitic wasp, of which we got a few dozen specimens in traps with rotting meat (originally intended for carrion beetles and flies), that he could not match with any known species. It belonged to the genus *Aphaereta*; these are wasps that lay their eggs in the pupae of bluebottles, which explains why it came to the meat-baited traps. In the next few days, the neighborhood naturalists worked with Van Achterberg on a scientific publication in which the species was described and named *Aphaereta vondelparkensis*. Until it was found elsewhere, the new species would be considered endemic to the Vondelpark, a great way for the local conservationists to highlight the ecological value of their park.

The lightning speed at which this new wasp was discovered, named, and described already answers one of those common questions: no, there is no international committee that has to vet the discovery before you are allowed to go public with it. You can just go ahead and publish your new species in any scientific journal. There *is* an international committee, however, that sets the standards that such a publication needs to adhere to. For example, you need to give a description in which the differences with known similar species are spelled out; you need to designate one specimen as the holotype, the species's reference specimen, which has to be placed in a public museum; and you need to give it a scientific name that is unique and does not go against "good taste" (one of the reasons why you have to be careful with naming a species after a person, as one person's hero is another one's villain). Other than that, anybody can publish a new species. It happens about 20,000 times per year.

Of course, it could be that later, another scientist ascertains that what you thought was a new species is actually identical to a species that had been discovered and named earlier. In that case, the rule is that the oldest name gets priority and the younger name is no longer used (it is dissed as a "junior synonym"). Also, species are often shuttled around from one genus to another. For example, our wasp species *vondelparkensis* was placed by us in the (already existing) genus of *Aphaereta*—a larger group that contains multiple species. But it could be that it is later found not to belong in *Aphaereta* after all but in another genus, say, *Chorebus*. In that case, the species name *vondelparkensis* would simply be taken out of *Aphaereta* and placed into *Chorebus*, so our Vondelpark wasp would henceforth be known as *Chorebus vondelparkensis*.

On the subject of scientific names, you may wonder why we still use those antiquated denominations, now that there are good English names for so many organisms. The reason is that the official biological naming systems recognize only those Latin names, not the English ones. No matter how many people call a house sparrow a house sparrow and never *Passer domesticus*, the only unambiguous name (the name that you would find in GenBank, linked to the house sparrow genome, for example) is *Passer domesticus*. One of the reasons for that is that English is not the only language in the world. If we did not have a unified, international scientific name for this bird, how would we know that a scientific paper in Japanese (イエスズメ), French (*moineau domestique*), and Peruvian Spanish (*gorrión casero*) all refer to the same animal? Another reason for using scientific names is that many organisms simply do not have vernacular names. Almost none of the hundreds of species of parasitic wasps that live in Amsterdam have a common Dutch (or English) name, so referring to *Aphaereta vondelparkensis* as the "Vondelpark wasp" sounds nice but does not carry much meaning if none of the other species has a common name.

Latin names were, of course, invented at a time when Latin was still the lingua franca of scholars and was widely known and understood, in the same way that English today is the international language of science. I agree that many people find those latinized names hard to remember, pronounce, and understand, and prefer to use common names, especially if one wants to communicate science to laypeople. Still, for serious naturalists, professional and amateur alike, in the end there is no other option but to memorize those scientific names as well: it is the only way to communicate about species unambiguously and across borders. Perhaps at some point in the future, some sort of large-scale overhaul will be possible, with the replacement, perhaps overnight and by algorithm, of all Latin names by easily memorized English equivalents that would henceforth have the same unwavering status. But until that time, we are stuck with those quaint scientific names, for which, in case you are wondering, the convention is that they always need to be underlined or *italicized* in a text.

Daunting as the naming system of organisms may seem to some urban naturalists, discovering new species is still one of the purest types of urban exploration I can think of, next to the sleuthing of secret ecosystems like the urban karst and microbial mats that we saw in the previous chapter. (Those elusive ecosystems, of course, will also yield many new species, like that pink subterranean eel and probably the majority of the species that DNA analysis uncovered in the microbial mats of Paris gutters and Rio de Janeiro church façades.)

Some urban discoveries of new species are accidental: widespread species that live outside cities too but just happen to be found for the first time in an urban setting. That was probably the case with our wasp in the Vondelpark, and also with the wasp *Anisopteromalus quinarius* that the Russian entomologist Alexander Timokhov discovered in his own Moscow apartment, where it was parasitizing beetle larvae that were infesting stored food in his pantry. But other species are discovered thanks to the special urban ecosystem that they inhabit.

For example, the ant *Strumigenys ananeotes,* which happened to have chosen to make its nest in the backyard of one of the world's prime ant specialists, John Longino, a myrmecologist who has published nearly 150 papers on ants and on his unashamedly ant-oriented résumé lists no fewer than ten ant species named after him (think *Megalomyrmex longinoi* and *Aenictus jacki*).

Longino's home is in the Avenues, an old residential area in the northeast of Salt Lake City, Utah. One evening in August 2018 he was rummaging around in his garden when he spotted a couple of *Strumigenys* ants walking around the potting soil. That is odd, Longino thought to himself, because *Strumigenys* is a kind of ant that lives in warm moist forest, where it hunts for springtails. Western North America is normally too dry for them, and only a few species are known from wet riverside habitats in California and Arizona. In Utah, they had never been seen before, so Longino suspected it must be one of those *Strumigenys* species that are transported with potting soil and plants all over the world—*S. eggersi,* for example, which is so ubiquitous that it deserves its moniker of "tramp ant."

Just to make sure, the next evening, Longino dug up a bucketful of soil from his garden and sifted through it carefully, coming up with sixty-six workers, a couple of larvae, and six winged queens. But when he began microscopically examining them, he was surprised that it was not at all *S. eggersi* or any of the other invasive species. Instead, it was a completely new species (which he named *Strumatogenys ananeotes* nov. spec., meaning "newly emerged *Strumatogenys*"), closely related to two species endemic to nearby Arizona and therefore probably a local, indigenous species, never recorded there before. But why would such a special species turn up in his freshly watered flower beds?

As he and his colleague Douglas Booher explain in their 2019 paper in the *Western North American Naturalist*, the reason probably is precisely that freshly watered flower bed, sitting in the green and wooded Salt Lake City, a lush, artificial oasis in the middle of a landscape where the natural vegetation is treeless grassland. "This new species is likely a relict," they write. Its natural habitat is probably subterranean moist sites where it had escaped detection by entomologists. "But human activity has created what is effectively a temperate broadleaf deciduous forest. The summer irrigation creates many square kilometers . . . with warm, moist leaf litter and garden soil. This species may be experiencing a Renaissance, reemerging after a long retreat belowground. In this case, the creation of an urban forest was not a disruption or displacement of its native habitat, but an expansion."

Other new species discovered in cities are of a similarly relictary nature. Remember that I said that vertebrates are so well studied that they are unlikely animals to yield new species in cities? Well, not entirely: on April 19, 2018, Kai Wang of the Kunming Institute of Zoology discovered *Lycodon obvelatus*, a new species of wolf snake (so named because of its canine-like fangs), during a nighttime foray in a city park in Panzhihua, Sichuan, China. He found the two-foot-long nonvenomous snake with its pretty cross-banding hunting for geckos on a stone parapet in the urban park.

The more or less circular park of about one kilometer across is an island of green in the center of this city of 1.2 million inhabitants built against the steep slopes of the Jinsha River. The mountain forests around Panzhihua were clear-felled in the middle of the twentieth century, and this park is the only surviving remnant. Even if it is filled with roads, playgrounds, tourist infrastructure, and exotic ornamental plants that are kept pest-free by frequent dousing in pesticides, it is probably the last stand of the snake's original habitat. The new species's name, *obvelatus*, says Wang, means "concealed," referring to the fact that "new species can be hidden even in major urban areas."

Tiny isolated pockets of habitat like the square kilometer of Sichuan mountain forest surrounded by urban sprawl where the last recluse wolf snakes live are typical environments in which to explore for new species. And, as we saw earlier in this chapter, so are those enigmatic habitats like the urban karst and microbial mats. Other urban ecosystems that probably hold many surprises for the urban naturalist who is hoping to discover new or rare species are typical extreme urban habitats. Think of bizarre, city-bound human-made ecosystems such as architectural green façades, polluted industrial sites, or the freshwater fauna living in the warm cooling water of power stations.

But there are more hidden treasures waiting to be revealed in the urban ecosystem, perhaps not so tangible as entirely new species of snakes or previously unknown microhabitats but at least as meaningful for understanding the way urban nature functions. For example, the biodiversity patterns created by the insular patchwork of green and gray that is a city.

10

URBAN ISLANDS

Ranjith De Silva, my colleague at Universiti Malaysia Sabah, in Malaysian Borneo, ran a seaweed farming project at Pulau Banggi, a large island at the northern tip of Borneo. There he commandeered a small flotilla of boats and sailors that could take him anywhere in this remote area. After begging him for months, I was finally invited to join him on his boat for a few days so that I could sample land snails on some of the tiny, otherwise unreachable islands in the shallow sea between Borneo and the Philippine island of Palawan.

To the field biologist, islands are irresistible. When you step onto a small oceanic island, far from anywhere, you enter a bite-sized chunk of nature, isolated from the outside world, where the forces of ecology and evolution have crafted unique, local kinds of life. They are places where evolution has built entire ecosystems from whatever biological flotsam the wind and the waves have brought in, home-grown food webs woven from variations on the few themes offered by a couple of random colonists. Places, too, where fragile species and crazy modes of life persist simply because they are untouched by the continental forces of relentless competition and ruthless predation.

The great science writer David Quammen in his wonderful book *The Song of the Dodo* has waxed more lyrical about the lure of islands than I could ever manage, so I will use his words to explain why I was so keen to accompany Ranjith: "Many of the world's gaudiest life forms, both plant and animal, occur on islands. There are giants, dwarfs, crossover artists, nonconformists of every sort. These improbable creatures inhabit the outlands, the detached and remote zones of landscape and imaginability. In fact, they give vivid biological definition to the very word 'outlandish.'"

Indeed, I discovered several locally evolved snail species in those three days in the Sulu Sea with Ranjith—endemic species that live nowhere else on Earth but on the islands where their ancestors landed and where they evolved into something truly local and unique. But there was another reason why I was so keen to go there: the species-area relationship.

Even though small tropical islands off the coast of Borneo are fascinating and contain unique species, their overall biodiversity is poor. On Lungisan, which is little more than a frigatebird-infested rock in the middle of the sea, you find only ten different snail species. A few tens of kilometers away, in a forest on the mainland of Borneo, you can easily find over fifty species of snail crawling around in an area the size of Lungisan. As Quammen writes, "Not only are islands impoverished relative to the mainlands, but small islands are more severely impoverished than large ones. That bit of insight became famed as the species-area relationship."

The species-area relationship has been discovered several times throughout the history of ecology, but most famously and most definitively in the 1960s by entomologist Edward O. Wilson (whom we met earlier in this book, when he talked about his "bug period" that stuck) and theoretical ecologist Robert MacArthur. They realized that the natural processes that determine which species live in a certain place are different on an island. Think of a newly formed island, the three-square-kilometer volcanic island of Surtsey, for example, south of Iceland, which rose up from the sea unexpectedly on November 14, 1963.

One and a half years later, when Surtsey had cooled down sufficiently for plants and animals not to be immediately incinerated, the Icelandic authorities had the good sense to found the Surtsey Research Foundation, which was to study how life settled there over time. And settle it did—slowly but surely. That year the explorer Sturla Friðriksson found a few individuals of the Arctic sea rocket, *Cakile arctica*, but nothing else. "When walking along the coast of the virgin island," he writes, "it was an amazing sight to see the first leaves of a sprouting seedling like a small green star on the black basaltic sand." *Cakile arctica* stayed for three years but disappeared and reappeared repeatedly in the decades since Surtsey's birth. In 1966 lyme grass came, which was rare for twenty years but in the mid-1980s suddenly boomed. Sea mayweed came in 1967, disappeared again the next year, and then returned for good in 1972. Also in 1967 came the first plants of sea sandwort, *Honkenya peploides*, which soon grew in its hundreds of thousands. And so, in fits and starts, the vegetation built up. From one plant species in 1965 to two in 1966, four in the years 1967–1970, and then six in 1971, it reached eleven species in 1972. But then stagnation hit, and for fifteen years the number of species hovered around twelve or so. Apparently, some sort of equilibrium had been reached.

And such an equilibrium is exactly what is predicted by Wilson's and MacArthur's theory of island biogeography. But not just any old equilibrium: it is a dynamic one. Surtsey's biodiversity may have remained stable at around twelve species, but the combination of species differed from year to year: in 1972, common chickweed went extinct and was replaced the following year by red fescue. In the early 1980s, fragile fern was gone, but smooth meadowgrass came. Of the eleven species of plant that grew on Surtsey in both 1972 and 1984, only seven were the same.

What Wilson and MacArthur discovered is that the biodiversity of an island, any island, is a balance between new species arriving and established species going extinct. When the island is fresh, every arrival is new, so the biodiversity grows. But as more species settle, the chance that a species goes extinct also goes up, simply because there are more species that might be hit by a random catastrophe. At some point in time, the number of species going extinct per year becomes (roughly) the same as the number of new ones arriving, and an equilibrium biodiversity is reached, even though the identities of those species keep changing all the time.

And that is not all. The biodiversity at which this equilibrium is reached is not the same for all islands. A small island "catches" fewer immigrants, and the immigrants that do settle reach smaller populations, so they are more vulnerable to extinction. Also, a small island has fewer distinct habitats, so there is less opportunity for a species to find its favorite niche. The outcome is that smaller islands harbor fewer species than larger ones. And another important thing is distance to the mainland: fewer species land on faraway islands than on islands close to the mainland, while their extinction risk is the same, so the biodiversity of remote islands will also be lower.

Over long times and large spaces, even such haphazard processes as colonization and extinction become regular laws of nature and, as Wilson and MacArthur deduced, transform into the mathematical laws of island biogeography. For example, when my students and I plotted the numbers of snail species that we found on those fourteen islands off the coast of Borneo, we obtained a curve that rose with island size and answered to the algebraic function of the logarithm of the number of species = 0.17 × the logarithm of the island's surface area + 1.26.

Now, I do not mean to scare you with this bit of mathematics, but I do want you to know that island biogeography is very important. It marries the romance of exploring remote verdant islands in a deep blue sea with the exact, black-and-white, numbers-and-graphs side of ecology. Moreover, it also shows us that there is a certain predictability in the study of biodiversity. Our project on snails from islands and the formula we derived from it mean that if somebody were to ask us about the snail biodiversity of any island off the coast of Borneo, even ones we had never visited, we could give a very accurate prediction based only on the island's size and its distance from the coast, without seeing a single snail shell from there.

But what do island biogeography, ecological formulas, and the jungle-covered islets that pucker the horizon of the seas around the north cape of Borneo have to do with community science in urban settings? As it turns out, the theory works just as well in the green archipelago of the city.

ANTHROPOCENE ARCHIPELAGOES

People often do not believe me when I tell them that I am just as happy and excited sampling the biodiversity of small, urban neighborhood parks as of those raw, remote, emerald islands of the tropical seas. And yet it is true: the feeling I had when I first entered De Slatuinen, a secluded greenspace in the west of Amsterdam, was very similar to what I felt when I first set foot on Lungisan, twenty kilometers off the Borneo coast.

Superficially, the two settings could not be more different. Lungisan is a conical rocky island one hundred meters wide and one hundred thirty meters long; it has existed for at least ten thousand years; it has the unforgiving deep blue waves of the South China Sea crashing into the steep limestone escarpment that surrounds it, and my students and I could only land our boat on a small beach at the bottom of a steep slope on its sheltered southern side. It is covered by a tropical forest composed of the few tree and vine species that can survive in the onslaught of guano from the thousands of frigatebirds that descend here from the huge avian maelstrom that each night positions itself over the island.

De Slatuinen, only a little bit smaller than Lungisan, is an urban park created in the early 1990s. Surrounded by four-story-high buildings, it lies in a sea of noisy traffic. You can access it only via a wooden gate in the wall next to one of the buildings. Inside there is a tangle of maple, elder, ivy, and honeysuckle and the screaming of the exotic ring-neck parakeets that roost in the trees every evening.

And yet, despite the vastly different settings, both are, in an ecological sense, islands. Lungisan is one that embodies the traditional meaning of the word: a patch of land surrounded by sea. De Slatuinen is an insular patch of vegetation surrounded by an ocean of buildings, streets, and traffic. Like the animals and plants that live on Lungisan, those in De Slatuinen also had to reach it by crossing a hostile terrain, and form their own little isolated ecosystem. Not just land surrounded by sea answers to the ecological definition of islands but any patch of habitat encircled by a barrier is an ecological island to the organisms that live there. Lakes, sinkholes, mountaintops, and the rocks poking from the ice that the Inuit call nunataks, to name but a few, are all considered islands. And, yes, so are urban greenspaces.

The municipality of Amsterdam has published a map of all the parks and other public greenspaces in the city, which stand out as an archipelago of patches of bright green among a sea of gray built-up areas. You see tiny ones, like De Slatuinen, medium-size ones, like the ancient Jewish Cemetery, and large ones, like the Vondelpark, where we discovered that new parasitic wasp of the previous chapter. Some are on the periphery and, like islands close to the mainland, are kept well stocked by the plants, animals, and fungi that live in large nature reserves in the countryside nearby. Others lie embedded in the center of the city, isolated from reservoirs of biodiversity by distance, buildings, and infrastructure, and are as insular and have as impoverished a biodiversity as remote islands in the middle of the ocean. Some are young, like the Diemerpark, which was founded in 2003. Others are much older, like the Keurtuinen, enclosed courtyards of the seventeenth-century inner city.

Sampling and studying the biodiversity in these urban islands, which we did during six summers with teams composed of people from the local neighborhoods, was as exciting and revealing as going to those tropical islands off Borneo. In De Slatuinen we found, as island biogeography predicts, a smaller number of species than in the much larger Flevopark, and the biodiversity of the even more extensive Diemerpark was larger still.

Amsterdam is, of course, by no means unique. All cities are perfect testing grounds for island biogeography. Grab a map of your town or city and pick, say, ten parks of varying sizes. Explore their biodiversity for birds, plants, mushrooms, or any other kind of organism, plot your numbers on a graph, and you will be able to see the patterns predicted by Wilson's and MacArthur's famous theory emerge in front of your eyes. You can also go all out, as a team of Chinese researchers led by Zhiwen Gao did for Kunming. In this sprawling city of four million, they drew eight imaginary lines ("transects," ecologists call them) from the center radiating out to the suburbs like the legs of a spider. Along each of the transects, they picked patches of vegetation at regular intervals and sampled the flora there. In 190 patches, ranging in size across four orders of magnitude, from 128 square meters to nearly 400,000 square meters, they found a curve of a steady increase of plant richness with increasing patch size, nicely following the island biogeography rule. They also discovered that patches close to the city center had lower biodiversity than similar-sized patches close to the perimeter and thus closer to the "mainland" of forests and meadows that could supply fresh colonist plants.

But the patchiness that is such a hallmark of cities also means that there are more ways to take an island biogeographer's view of the urban ecosystem. In the UK, then biology teacher Alvin Helden and ecologist Simon Leather, for example, took almost as many risks as a biologist landing her rubber dingy among crashing waves on a rocky islet in the sea when they ran the gauntlet of heavy traffic on the traffic circles (which in the UK are called "roundabouts") of their hometown, the city of Bracknell.

When you look at the map of Bracknell, you understand why. All the major thoroughfares in the city are interrupted by traffic circles. Some are small, like the intersection connecting the three spurs of Doncastle Road; others are huge, such as the one of more than one hundred meters across that links the A329 with London Road. More important, poring over that map also reveals that most of these traffic circles have a green inner core: a perfectly round minipark with grass, bushes, flower beds, and trees.

Helden, who has a lifelong interest in Hemiptera, the insect group that includes bugs, cicadas, leafhoppers, and aphids, and Leather, an ecologist at Imperial College, London, saw the opportunity that this offered. Even more than urban parks, these rotaries were, to all an ecologist's intents and purposes, islands: road islands of vegetation floating in a sea of raging traffic. Leather recalls: "To get to work every day, I had to negotiate thirteen roundabouts of varying sizes and appearance and it suddenly struck me that here was a perfect way to talk about urban ecology, island biogeography and nature conservation."

So the two approached the city council, which was "remarkably laid back about the whole thing," donned bright yellow safety jackets, and threw themselves into the circling traffic. They took to the road islands' vegetation with inverted leaf blowers to suck up small bugs from the grass and inverted umbrellas to catch the bugs they shook from overhanging branches. And, once again, perfect island biogeographical patterns were revealed in the numbers of species of bugs that they found.

Parks, road islands, road verges, even the dynamic world of inner-city vacant lots that come and go like an urban version of a volcanic archipelago are perfect urban playgrounds for the itinerant island biogeographer. But the fragmentary nature of urban habitats offers more opportunities to the urban naturalist than just confirming Wilson's and MacArthur's laws: the city's green archipelago turns out to be a fascinating patchwork of environmental idiosyncracies, with each fragment its own little world.

ISLAND GARDENS

For example, when mite expert Henk Siepel and I began studying the soil mites of the Keurtuinen, we discovered that these old enclosed courtyards in the center of Amsterdam are like time capsules. The soil mites there form an assemblage of species found nowhere else in the city today. They are probably an ecological relic from the seventeenth century, when the city expanded into the rural hinterland and parts of ancient, natural habitat were enclosed inside blocks of five-story-high canal houses. Since then, airborne species (flying insects, birds, plants, and fungi dispersed by seeds and spores) would have been exchanged with the outside world, but small flightless organisms such as soil mites would have stayed put, unable to keep up with the ecological changes that took place in the rest of the city.

Further east in the city, the old Jewish Cemetery, its sacred soil undisturbed for centuries, also harbored a fauna much more original compared with that in the neighboring Flevopark with its dogs, playgrounds, jogging tracks, and lawns for picnicking.

Another unexpected insight into the biodiversity of the urban green archipelago was revealed when my graduate student Maaike de Voogd started looking at urban gardens. She placed pitfall traps in ten backyards in each of three Dutch cities. Simultaneously, she did the same in ten sites in each of three nature reserves close to each of the three cities. Two weeks later, she picked all traps up again, and spent almost a year identifying and counting all the little soil animals that the traps had caught: snails, beetles, millipedes, centipedes, woodlice, and harvestmen—in total nearly five thousand of them, belonging to 360 different species. Crunching all the numbers, she made a startling discovery.

To explain what she discovered and why this is important, we need to talk about three different levels of biodiversity, called alpha, beta, and gamma diversity. The first, alpha, is simply the number of species you find in one single spot—in De Voogd's case, in a single garden or a garden-sized patch of nature reserve. The last one, gamma diversity, is the total diversity you find in all the gardens of a city taken together, or in the entire nature reserve. And the measure in between, beta diversity, is what connects alpha and gamma: the changes in the lists of species as you move from garden to garden or from one site in the nature reserve to the next.

What De Voogd discovered is, perhaps not surprisingly, that alpha diversity in gardens is relatively low: she usually found only a little over a dozen species in a single garden compared with a doubly as rich biodiversity in the nearby nature reserve. The gamma diversity is also lower in city gardens, but not *that* much lower: across all three cities, the total gamma biodiversity of the gardens was around 250 species, while that of the nature reserves was 320 species. But the most exciting finding was what happened at the level of beta diversity: while she would encounter pretty much the same species of soil animals as she moved from one site to the next in her nature reserves, the garden fauna between neighboring gardens was strikingly different. In one garden she would often find quite a different set of creepy-crawlies than in the garden of the neighbors.

De Voogd thinks this is because the backyards are governed by different ecological processes than forests and other nature reserves. In the latter, large-scale things such as climate and soil type would be the main determinants of what kind of environment you would find in a place, so two adjacent plots in a nature reserve would probably be very similar. But in a garden, the ecosystem is under the firm dictatorship of the person who tends it. In a street, two neighbors might have vastly different opinions on garden design.

At one house, the residents might be enlightened souls who planted their garden with lots of native species of herb, tree, and shrub. They might have a compost heap and a pile of decaying logs, and in the corner a pond with a gently sloping bank and lots of aquatic and swamp plants. In that garden, De Voogd would find a rich assemblage of shade and moisture-loving species akin to what she would catch in her forest pitfalls. But the next-door neighbors might adhere to the kind of conservative, unimaginative tradition of laying down tiles and pebbles and planting a row of low-maintenance conifers in black soil covered in wood chips. The biodiversity there would consist of a smaller and completely different set of heat- and drought-tolerant species inhabiting what little space was left between the wood chips and the pavement tiles.

So, even though the alpha and gamma diversity of backyards may be low, the combined (beta) biodiversity of several small gardens, with its gardeners of very different temperaments, and consequently very different mini-ecosystems, would probably be much greater than the (alpha) biodiversity of a single large one. And that brings us to a long-standing debate in conservation, which can be applied to urban conservation as well: Single large or several small?

SLOSS

Simon Leather, whom we saw sampling bugs on road islands in Bracknell in the previous section, said that his project on the biodiversity of such seemingly bland and uninteresting localities as traffic rotaries was vital "to introduce students to practical research which might actually lead on to a career in ecology or nature conservation." The reason is that urban island biogeography touches the core of a decades-old but still raging debate nicknamed SLoSS, which stands for "single large or several small."

When faced with hard choices about which ecosystems to preserve in an already fragmented area, conservationists often need to decide between conserving a lot of small fragments or one larger area. Think, for example, of cement companies quarrying away limestone hills: if quarrying must go ahead, should we set aside one large hill for conservation and leave the smaller ones for the cement company to eat up with their digging machines? Or rather the reverse? To the cement company, it may not make much of a difference: limestone is limestone. But for conservation, the choice is crucial, because the ecosystem on one hill may not be replaceable with that on another hill.

Proponents of "several small" say that although island biogeography teaches us that the biodiversity in each of those small fragments will be low, it also teaches us that each island will harbor different sets of species, so the accumulated biodiversity encapsulated by several small reserves will be high. This is what Maaike de Voogd discovered with the several small city gardens of the previous section. But proponents of the other camp point out that the biodiversity in the single large reserve may be even higher than that of the smaller fragments combined, and moreover that it will be much more robust, since the populations there will be larger and contain more genetic variation; hence the reserve would be expected to be more resilient in the face of disease, drought, or other ecological disasters. Also, a single large reserve might be easier to defend against the onslaught of human or natural forces of attrition.

Over the years, urban nature conservation practice seems to have come down on the side of single large. Or at least there has been a tendency to connect the many small fragments that cities inevitably generate with green corridors for animals and plants and essentially build a dispersed, connected, single large urban nature reserve. For example, the city of Beijing, which over the years has been expanding like an archery target with ever more concentric rings into the 22 million-inhabitant megalopolis that it is today, has been developing a network of green corridors. The aim is to "liberate" the many isolated parks and other insular greenspaces locked inside those rings and the ring roads around them. The vegetated corridors, some newly designed and built, others developed by making use of derelict railway lines and road embankments, are meant to save species marooned in green islands from extinction and prevent such urban islands from ecological collapse. Many other cities have done the same, and have developed urban ecology policies largely built on the premise that, in cities, connectedness of greenspaces is key to their survival, and SL is better than SS.

Of course, green corridors in cities are a good thing, if only because they add to the total amount of greenspace. But there are reasons to cast doubt on the dogma of connecting everything that is isolated. As we saw above, some fragments of urban nature are unique because of the way they have been managed or because they are ancient and preserve remnants of long gone ecosystems. The fact that they are isolated fragments is key to their survival. If we were to open up those unique habitats to the outside world they might be swamped by the more ubiquitous urban flora and fauna of generalists, invasive species, and strong competitors, and suffer a demise similar to what has happened to fragile ecosystems on oceanic islands after the introduction of species from the mainland.

In the scientific literature too, cracks have begun to appear in the foundation under the idea that single large is better. Lenore Fahrig of Carleton University in Ottawa, Canada, trawled the ecological journals for articles on the effects on biodiversity of fragmentation per se—that is, if a habitat is carved up into several patches without the total surface area being reduced, how does biodiversity respond to this? As it turns out, of the almost four hundred cases she reviewed, over three quarters showed, contrary to received wisdom, a positive effect on biodiversity. And that positive effect, probably caused by many different factors, including the ones I mentioned in the previous paragraphs (and more that we will meet in a later chapter), held up across all different types of organisms and also was the same for common, widespread species and rare, sensitive species. For example, one study in three cities in Switzerland discovered that isolated greenspaces were usually more different in character and had an overall higher biodiversity for arthropods. At the same time, the fact that these patches were isolated did not seem to affect the animals too much because these urban species are very good at moving around in the harsh urban environment.

In fact, the results from her study were so strong and robust, and had been found and reported in scientific articles so many times and for so long, that Fahrig called the idea (that fragmentation should be bad) a "zombie idea": an idea that should be dead but is not.

Its undeadness was demonstrated as soon as Fahrig's paper appeared. A defensive force of seventeen ecologists from all over the world rose up from those branches of ecology and conservation that are in the business of improving the connectivity between small nature reserves and, in a paper in *Conservation Biology*, accused Fahrig of cherry-picking. They thought she had selectively chosen examples from the literature that supported her idea. They also berated her for playing right into the hands of those evil forces that would do nothing rather than fragment habitats.

It did not take long for Fahrig to strike back. She also assembled a crew of over twenty co-authors for a rebuttal in *Conservation Biology* in which they effectively said no, our analysis stands, there was no cherry-picking, and the risk of scientific results being misconstrued and abused should never be a reason not to do the research.

And that is where the SLoSS debate stands. Somehow the zombie idea that fragmentation per se is always bad keeps rising up, even though there are many reasons to believe that small fragments have great conservation value. Fahrig and her team write, "What we find 'alarming' for conservation is the nearly complete lack of protection for habitat that is divided into small patches."

In the cities where I have worked so far, I have noticed the same sentiments. Small greenspaces like vacant lots, road islands, the overgrown edges of a roadside verge, or a tiny pocket park are often dismissed as of negligible importance for urban biodiversity. True, they might be too small to harbor a viable population of large birds or mammals. But they usually are large enough to provide an environment for an entire ecosystem of plants, fungi, and invertebrate animals, even when they are surrounded by traffic thoroughfares that condemn them to eternal urban isolation.

Speaking of traffic, that is where we will head in the next chapter. For in cities, roads and their users are one of the chief knives by which habitats are carved into smaller pieces. Like rivers of death, roads meander and encircle and disconnect smaller and smaller bits of greenspace. And regardless of what we talked about in this chapter—the value of small fragments and the interesting ways in which some aspects of biodiversity can accumulate when larger habitats are split into artificial archipelagoes—this does not do away with the fact that roads will wreak havoc on many animals as they cross them or even come near them. In ecology, nothing is ever simple: yes, the laws of island biogeography dictate that fragmentation begets diversity. But the laws of traffic also dictate that the price for this diversity is paid in lives lost on the tarmac. Death by traffic, that uniquely urban cause of mortality, is what we will be delving into in the next chapter.

JONO24

11

INVOLUNTARY SLAUGHTER

We saw it happening, my then girlfriend and I: the little bird suddenly flew out of the bushes by the roadside, darted low across the tarmac, and then disappeared into the right rear wheel of the car in front of us. It was centrifuged for a few revolutions before it was ejected sideways and landed, flapping in an uncoordinated fashion, in the verge.

I quickly parked the car, walked back, squatted down, and picked up the animal: a Sardinian warbler, a pretty, very common bird; petite and not at all resistant to the racking it had just suffered. My girlfriend squatted beside me. Together we watched how the bird attempted a few half-hearted wing flaps, weakly pecked my fingertip, and then lost the light in its little beady eyes and died in the palm of my hand.

At that moment, to my surprise as much as to my girlfriend's, I was overcome by sadness, and for a few minutes I sat there, crying, by the side of a country road in the Peloponnese, cradling a dead warbler in my hand, with claxoning cars swishing by and my girlfriend comforting me with a slightly bewildered look on her face. Fifteen minutes later, we were driving again, she quietly asked me how come the death of a bird affected me so, when I am such an animal mass murderer myself.

She hit the nail—or rather, the insect pin—on the head. Throughout my life as a scientist I have been responsible for the scientifically sanctioned deaths of hundreds of thousands of animals, mostly arthropods and mollusks. In fact, earlier that day I had merrily stuffed some snails into a jar of alcohol. And although I have never killed any vertebrates, I have regularly participated in field trips where others were collecting frogs, small mammals, and also cute little birdies like that Sardinian warbler, and never shed a tear.

So why would I cry over the death of this bird? Analyzing my emotions, I concluded that what had touched me was the utter senselessness of this death. An animal that is killed and preserved by a researcher contributes to the knowledge that we have of its species. It is lovingly curated, its features are recorded, it is the object of study and the subject of scientific publications, and it is preserved for eternity in a natural history museum collection. Yes, its life has been lost, but its body has obtained a new kind of value, a point I made in a previous chapter.

Roadkill is the complete opposite of that. That motorist did not kill that warbler intentionally; in fact, he or she probably never even noticed the collision. And that ignorance and lack of intent are what make the event so tragic. A little life has been ripped from this Earth (and, who knows, if it was a nesting bird with dependent chicks, several lives) and its value has been lost forever.

Not that the worth of these animals is obvious to everyone. The American anthropologist Jane Desmond, who made a study of the cultural impact of roadkill, concludes, "These animal lives have little value for most of the [people] in the United States, as these animals are unowned, lacking in monetary or emotional value, not pets or livestock, and without the charismatic following that megafauna like elephants and lions in zoos receive."

Nonetheless, most people are uncomfortably aware of our indifference toward the multitude of road deaths of pet-sized mammals and birds, and their unease emerges as humor. The faux field guide *Flattened Fauna* by Roger Knutson, for example, provides species-specific diagrams to recognize the two-dimensional shapes of pancaked animals and a handy "death list" in the back to tick off the ones you have spotted. Then there is Buck Peterson's trilogy of hillbilly-esque humor, the *Original Roadkill Cookbook*, the *International Roadkill Cookbook*, and the *Roadkill USA Coloring and Activity Book*. And, admittedly, when I lived in Malaysia and on the road passed through the unmistakable pong of a dead Sunda stink badger (*Mydaus javanensis*), would I not spontaneously burst into singing Loudon Wainwright III's all-time favorite tune, "Dead Skunk in the Middle of the Road (Stinking to High Heaven)"?

Crude humor aside, I recognize familiar sentiments in the work by artists who are, for lack of a better word, inspired by roadkill. In her *Roadside*

Memorial Project, the Kentucky-based artist and activist L. A. Watson has been creating reflective silhouettes of animals, which she installs among the grass in the verges of roads in her home state wherever an animal of that species has been run over. They function both as a warning sign to drivers and as a memorial to the dead individual in question. On the project's home page, Watson writes: "The color white was chosen . . . because it references the iconography of human roadside memorial crosses and denotes innocence, sacrifice, spirits and ghostly specters. The installation comes to life at night, and is 'turned on' by the passing drivers who illuminate it, many of whom slow down."'

In a similar project, the eco-artist Brian D. Collier builds roadside shrines for animals, complete with the date of death and a color picture of the deceased, akin to the ones we sometimes see erected along the road for people who have lost their lives in traffic accidents. Jane Desmond writes of his work, "Streetside shrines . . . to roadkilled animals may be tiny acts of recognition, but they point to the possibility of greater emotional cognizance of animal carnage on highways. . . . Similar shrines . . . could clutter the road with the marking of accumulated deaths, too innumerable to count as we go whizzing by."

And clutter the road with shrines is exactly what my colleague Bram Koese did. Koese is one of the best freshwater zoologists of the Netherlands, specializing in mayflies and caddisflies but with a near encyclopedic knowledge of most other aquatic animals, and terrestrial ones, for that matter. He lives in a town, surrounded by wetlands and canals, some thirty kilometers south of Amsterdam, and takes regular bicycle rides along the Ziendeweg, a narrow road between his hometown and the next. During rush hour many commuters use it to circumvent the traffic jams on the highway. And these speeding cars often hit wildlife, Koese noticed. He saw entire families of graylag geese (*Anser anser*) and barn swallows (*Hirundo rustica*) being mowed down. Prompted by these sad encounters, and curious about the actual impact of the traffic on wildlife, he began logging his roadkill sightings on the citizen science platform Observation International. For a whole year, on average every other day, he would ride up and down the road, scanning with a headlamp if it was dark, and record and photograph every dead animal (birds, mammals, amphibians, even the occasional butterfly or migrating crayfish) and its location.

His sightings amounted to 642 carcasses. The "death list" included thirty-five mammals, ninety birds, and 515 amphibians, among which were rare and protected species such as the stoat (*Mustela erminea*), weasel (*Mustela nivalis*), European polecat (*Mustela putorius*), tawny owl (*Strix aluco*), moor frog (*Rana arvalis*), and natterjack toad (*Epidalea calamita*). Shocked by the volume of his data, and disenchanted by the lack of response he got from the municipality, Koese then hatched a secret plan for a clever guerrilla campaign.

Quietly asking around in the neighborhood and among family and friends, at the local nature conservation organization, and at the bicycle workshop, he managed to round up some twenty-five coconspirators. Together they assembled not one but 642 roadside shrines. Each was made from a one-meter-long stake of live willow wood (leftovers from recent willow coppicing—so they could still take root when stuck into the ground). Friends with a large garden sawed the crossbeams from old floorboards of Koese's stepfather's and whitened them, while Koese and his girlfriend whitened all vertical stakes with chalk (it took them twenty days). Each cross would come to represent a separate roadkill exactly at the spot where it had taken place. The name of the species and the date of its demise were stenciled on the crossbeam, and a picture of the live animal was placed next to a QR code that led to the record on the Observation International website, where a photo of the animal in its flattened state could be viewed.

Installing the crosses turned out to be quite an operation because the team wanted to eschew the use of any motorized road vehicles and also because the COVID-19 regulations required many online briefings to coordinate and discuss the marching orders. On Thursday, May 7, 2021, a friend of Koese's came with his flatbottom boat. With the help of five workers from the bicycle workshop, they transported all the stakes from Koese's house to the boat ("The guys found the work a little tedious, but we managed to motivate them with chocolate cake," says Koese) and then on to the Ziendeweg (which runs parallel to a canal). That evening, a team of three volunteers used chalk to mark all 642 locations on the road, while another team transported the stakes, by boat and carrier cycles, and hid them, with ground drills and sledge hammers, in the bushes at three locations.

The next morning, at the crack of dawn, ten two- to three-person teams quietly took their positions along the road, each responsible for the installation of some sixty crosses. Then, using wheelbarrows, a canoe, and the flatbottom boat, volunteers distributed the stakes, drills and hammers among the groups, while, still in the quiet predawn hour, two volunteers on carrier cycles handed out the crossbeams with the names and the QR codes. Once everything was in place, the teams quickly and simultaneously began hammering away and installing the crosses along the roadside. When the sun was up, and before the rush hour began, it was a magnificent sight, and looked exactly as Jane Desmond had envisaged it: an endless parade of white crosses stretching to the horizon along the four-kilometer-long straight road, making every driver (as well as the local authorities, who learned about it from the local and national media that Koese had notified) painfully aware of the "innumerable accumulated deaths."

But not for long. As it turned out, not everybody in the neighborhood was sympathetic toward the campaign, and within a day after they had been set up, all the crosses had been kicked down (perhaps by the same person who had stuffed a recently installed underpass intended for otter crossings with plastic foam and set it on fire). Koese and his team resurrected most of the crosses, only to find them again vandalized the next day. But they were not disheartened: by that time the project had served its purpose. Koese, his team, and their cause were in the news for the whole weekend, the local authorities had been presented both with a letter and a scientific report with a detailed analysis of the findings, and a strong case had been made for the road to be out of bounds for anything but local traffic.

Koese's project drives home two things. First, that road ecology, as it is called, is a perfect subject for community science projects. Good online platforms exist for logging roadkill events, *and* the community group could adopt guerrilla tactics for their campaign that an "official" project would probably not have got away with. (Moreover, these community projects tend to be focused on the impact of traffic on wildlife, whereas many officially approved roadkill monitoring projects are begun for the opposite reason: to control the impact of wildlife on traffic.) The second point is that one roadkill is a tragedy, but a million roadkills are not just a statistic. By upscaling from a single roadside shrine in memory of a single deceased animal to a mass grave that showed the actual scale of the problem, Koese and his team were able to make us seamlessly progress from mourning the loss of one individual animal's life to grasping the danger that entire populations are exposed to.

DR. SPLATT

In fact, most roadkill monitoring projects all over the world either are conceived by community groups or at least rely on citizen science observers. Some, like the award-winning RoadKill school study, which was thought up in 1992 by Brewster Bartlett (aka "Dr. Splatt"), a science teacher in New Hampshire, began long before citizen science, apps, and the internet were commonplace. Bartlett and his students logged roadkill seen on their commutes between home and school, and the school's very first email server was used to exchange information on sightings and to post the data to a bulletin board. Later, after CNN reported on it in 1997, other schools joined in, and Bartlett became something of a celebrity, giving lectures across the nation in his white Dr. Splatt lab coat and carrying around a box full of preserved animal bits picked up from the roadside. Dr. Splatt himself retired in 2018, but his project is still active and is now one of the longest-running citizen science projects out there.

Since Dr. Splatt blazed the trail, many other roadkill community projects have sprung up across the globe. In Belgium, which has Europe's densest road network, a joint governmental-NGO project has been running since 2012 to map the country's estimated six million roadkills of vertebrate animals per year. The project, with the rather confrontational moniker *Dieren Onder De Wielen* (Animals Under Your Wheels), allows drivers to use speech recognition on the app ObsMapp (a data upload app of Observation International) to report and log roadkill.

Also making use of Observation International is *Plan SAFE* (*Stop Atropellos Fauna de España*—Stop Abusing the Animals of Spain), and the Society for the Protection of Nature in Israel (SPNI) launched a roadkill mapping project that relies on a feature in the popular (and, originally, Israeli) navigation app Waze. Motorists can tap an icon (the face of a raccoon with crosses for eyes and its tongue sticking out) whenever they spot a dead animal on the road.

Thanks to these and many other projects around the world, we now have a pretty good handle on the statistics of roadkill. In 2020, Clara Grilo of the University of Aveiro in Portugal and her colleagues combined ninety European roadkill studies in a calculation that showed that on Europe's roads alone, around 194 million birds and 29 million mammals die yearly. In the United State, similar calculations show that close to 400 million vertebrates are killed by traffic each year. Extrapolate those numbers to the global situation and you come up with something around two billion birds, mammals, reptiles, and amphibians squashed, mangled, and otherwise struck dead by road traffic each year, or 5.5 million per day.

Those numbers seem incredibly high. But when you divide those 5.5 million deaths by the 36 million kilometers of paved road that we have in the world, you come up with one death per day per seven kilometers. And then, judging by the numbers of sad little bundles of fur and feathers littering the hard shoulders of roads in the five continents that I have traveled by car in, that figure seems reasonable, if not a tad low, to me. In fact, there are good reasons to believe that all those estimates, despite the many zeros, are more than a tad too low. First, not all killed animals die on the road: many are mortally wounded but manage to flee the scene of the collision and die some distance from the road, never to be found. Others die on impact, but that impact is so forceful that their dead bodies are catapulted far off the roadside verge and similarly are not included in roadkill surveys. And then there is the matter of "carcass persistence."

In their zombie-titled paper, "How Long Do the Dead Survive on the Road," published in the journal *PLoS ONE* in 2011, Sara Santos and her colleagues from the University of Évora, Portugal, examine a gory but crucial question that is central to all roadkill surveys: How long does a carcass from a roadkill hang around on the road? That is, for how long does it remain visible before it is carried off by a fox or a buzzard or some other scavenger, or disintegrates under the natural forces of decay and rain, or is pressed into the tarmac beyond recognition? They point out that most roadkill monitorings survey the same stretch of road once a week or even less frequently, laboring under the delusion that most carcasses stay on the road for several days, so that no new kills are missed, and not too many are double-counted. But, say Santos and colleagues, nobody really knew how long or short an animal carcass actually "survived" on the road.

So they decided to study it. For more than a year, along thirty-seven kilometers of road in southern Portugal, Santos and her team monitored roadkill day in and day out, and recorded in detail the type and position of each of the 4,447 dead animals they encountered. With their data, they could tell exactly how long a specific carcass remained before it disappeared. The answer was, not very long at all. More than half the carcasses were lost without a trace in less than twenty-four hours, and only about 5 percent remained on the surface after a week. Of course, some bulky carcasses, such as those of large carnivores and birds of prey, or armored ones, such as hedgehog carcasses, were visible for more than a week, but the overall message was that a roadkill cadaver is a pretty ephemeral thing.

The good news from the Portuguese study is that roadkill surveys are not likely to include many carcasses that have been doubly counted. But the bad news is that the reverse is much more likely: many roadkill carcasses come and go in between two rounds of weekly surveys, leading to an underestimation of the numbers. The true global total is therefore likely to be much higher than two billion. And remember that this number is only for vertebrate animals, chiefly the pettable ones with hair and fur that most motorists would feel somewhat sorry for. We know even less of the slaughter afflicted on the much more numerous invertebrate animals (snails and slugs, insects, spiders, millipedes, and so on), though some very persistent researchers have tried to obtain numbers for those as well—James Baxter-Gilbert of the Laurentian University in Sudbury, Canada, for example.

For two consecutive summers, he and a bunch of colleagues and students took daily walks along a two-kilometer stretch of road some sixty kilometers south of Sudbury and picked up all the squished insects they could see. All 117,674 of them! Extrapolating from these numbers in an article in the *Journal of Insect Conservation*, they come up with a total of many hundreds of billions of insect deaths from traffic each summer in North America. Flies alone would amount to some 187 billion.

But, mind-boggling as they are, even these figures could easily be way too low. The Canadian team picked up dead insects while walking by the side of the road. This means their sample was limited to insects that they could spot from standing height, probably one centimeter in size or more, while we know that insect species of that body size are actually a minority: 90 percent of species are much tinier. So, to get a more unbiased sample of insect roadkill, Dutch researcher Arnold van Vliet of Wageningen University used a different approach. He devised a citizen science project called SplashTeller, which involved 250 volunteer car drivers who, for six weeks, took a daily photograph of all the insect splats on their car license plates, recorded the distance traveled, and then thoroughly scrubbed the plate to start with a clean slate the next day. All told, there were 17,836 splats for 30,873 kilometers of traveled road. Now, a Dutch license plate is only a small fraction of the frontal surface of a car, and the distance traveled only a small part of all the kilometers traveled by Dutch drivers per year; and the Netherlands are only a small country. So, again doing extensive but reasonable extrapolations, Van Vliet came up with estimates that imply that, globally, per year, 228,000,000,000,000 (228 trillion) insects are killed by getting flattened against oncoming vehicles.

I'll stop with the astronomical figures there. But I think it is obvious that road traffic causes huge losses of animal life. Still, let us not be blindsided by numbers alone, for what really counts in ecology is the proportion of a population that is killed. In other words, what is the denominator?

RUBBER SNAKE

A bunch of jet-black feathers glued to the road, pointing in weirdly perpendicular directions. A soft lump of fur on the rumble strip, hairs wafting to and fro with every passing car. An insect with a snapped abdomen, lying upside down on the hot tarmac, its rustling wings still fluttering in futile attempts to get airborne again. A crushed snail, its pale innards neatly spread out like in a diagram from a dissection manual, fragments of its shattered shell strewn around. Each of these horrible deaths is a tragedy in itself, but in ecology, it is not so much the individual that counts but what those individual events mean for the population at large—the statistics rather than the tragedies. And that depends on a great many things, such as what proportion of the population is likely to live near roads, how likely those animals are to cross the road, and, if they do, how likely they are to be struck down.

For birds and mammals, we have some of that information, so it is possible, as Clara Grilo did, to work out the threat that roadkill poses for the species. She found that certain species, such as the European blackbird (*Turdus merula*), despite suffering huge roadkill (a whopping 35 million per year), have such large populations and such a high rate of reproduction that they are able to absorb the losses without noticeable traffic-induced declines in numbers (though they are declining from other causes, and roadkill can exacerbate those losses). But other species, such as the European hazel grouse (*Tetrastes bonasia*), the Brazilian maned wolf (*Chrysocyon brachyurus*), and the brown hyena (*Hyaena brunnea*) from Southern Africa, are likely to be driven (literally) to extinction mainly by road traffic in the next few decades.

For smaller animals we often do not know enough to make such detailed calculations, but we can make some educated guesses. Many ground-dwelling insects such as ground beetles (Carabidae) or rove beetles (Staphylinidae) actually do not like to cross a road. When they hit a road, they will rather walk parallel to it, along the road edge. The result is that roads do not kill these animals so much as carve up their ranges into many smaller-sized fragments.

And then there are animals against which car drivers themselves have stacked the deck. In the 1990s, when zoologist Paul Ashley did a roadkill study in Canada, he noticed that lizards were surprisingly often found squashed on the side of the road or on the line between two lanes, places that car wheels normally do not touch. Could it be that drivers were actually *aiming* for these animals? To investigate this disturbing possibility, in 2005 Ashley went to a toy shop and got himself a realistic-looking rubber snake and turtle. He also bought an invisible grease marker of the same width as the rubber reptiles, as a control to measure how often car wheels would normally hit an object on the middle line by chance. Then he went to the nearest highway, placed one of the objects on the central line, and quickly hid in the bushes. For each passing car, he wrote down whether it hit the object and the gender of the driver.

After having done this for two thousand passing cars, Ashley, who wrote an article about this study titled "Incidence of Intentional Vehicle–Reptile Collisions," had to conclude that some drivers definitely have something against reptiles: the invisible grease marker was hit thirteen times (so this was the number of hits that were completely unintentional). But the turtle was hit twenty-two times and the snake thirty-one times. Doing some calculations, Ashley was able to work out that around 3 percent of the drivers must have targeted the reptiles on purpose, especially if they (the drivers, not the decoy reptiles) were male. However, Ashley saw that a similar number of drivers (again, mainly males) instead tried to rescue the animal and remove it off the road. His paper does not mention whether in such situations he then emerged with the awkward task of explaining why he had pranked them with a toy snake while hiding in the bushes. Although the rescuers largely offset the effect of the reptile haters, Ashley's study did show that we cannot simply assume that roadkill is accidental. For certain animals and in some parts of the world, drivers may have it in for the innocent animals that venture onto the highway.

COMPLETELY SHATTERED

As human threats to wildlife go, traffic is a relative newcomer to the sizable portfolio. The first documented case—documented, that is, in the diary of the grandmother of a friend of *Flattened Fauna* author Roger Knutson—of an animal killed by traffic reputedly was a snapping turtle (*Chelydra serpentina*), snapped under the steel-rimmed wheel of a loaded wagon in North Dakota in 1897. We have come a long way since then. As early as 1920, the naturalist Joseph Grinnell noted that "if one were to estimate the entire mileage of such roads in the state [California], the mortality must mount into the hundreds and perhaps thousands every 24 hours."

A few years later, in 1925, a first stab at getting actual data was reported in a paper published in *Science*, thanks to a car game played by Lillian and Dayton Stoner on a road trip from their hometown, Iowa City, to the Iowa Lakeside Laboratory. "It occurred to us that an enumeration and actual count . . . would be of interest," Dayton Stoner writes. Though today this would not be considered sufficient justification for publication in the world's foremost scientific journal, it did produce the first ever roadkill quantification. We can only imagine Dayton driving their T-Ford along an unpaved road at a gentle forty kilometers per hours with Lilian, her scarf flapping in the wind, keeping score on a notepad. "And another garter snake!" they might have exclaimed in unison above the roar of the engine. In all, they spotted 225 fresh carcasses of twenty-nine species (among which no fewer than forty-three red-headed woodpeckers) along the thousand-kilometer return trip. "The death-dealing qualities of the motor car are making serious inroads on our native mammals, birds and other forms of animal life," Dayton Stoner presages. But it took until the 1950s for roadkill surveys to appear regularly in the scientific literature. And the first books with "road ecology" in the title were published only in the mid-1990s.

Today, road ecology is a field that is growing nearly as fast as the world's road network itself, and it includes much more than just studying the impact of roadkill. For roads also bring many other environmental ills. Think of the fragmentation of habitats, changes in the hydrology of an area, and the provision of easier access to an area for hunters, loggers, vacationers, and many other people on a mission to exploit the area in one way or another. On the plus side, in intensively managed agricultural areas, roadside verges are often the only places where wildflowers can grow. And as our obsession with building roads increases, the importance and impact of road ecological processes will only grow.

Nobody, not even Google, knows for sure how many kilometers of road exist in the world and how all those linear paved elements combine to create a road-dominated landscape. One good way to improve this knowledge is by turning the users of roads into mappers, and that is exactly what the project OpenStreetMap has done. A global community of seven million volunteer mappers create up-to-date information on the positions and status of roads in their neighborhood. In 2016, in a paper in *Science*, a team led by Pierre Ibisch of the Eberswalde University for Sustainable Development in Germany used OpenStreetMap data to create a high-density map of the global road network of 36 million kilometers and how it carves up the land into smaller fragments. They concluded that 80 percent of the world's land surface is still more than one kilometer away from the nearest road. That is, if you were to stick a pin into a random land point on the map, the chance that the nearest road is more than one kilometer away is 80 percent. But roads do divvy up that roadless land into no fewer than 600,000 patches worldwide, half of which are less than one square kilometer and only seven hundred or so are more than one hundred square kilometers. In other words, given what we know about the impact of roads on the living world, we live on a planet that is completely shattered into tiny road-encircled fragments. And our obsession with roads is only about to get worse, with a projected 25 million additional kilometers of road to be built by 2050.

Clearly, in an urbanized world with an increasing and increasingly mobile human population, roads and other channels by which people are transported on land are a necessity, but their impact on the rest of the living world is becoming unacceptably large. What can we do to mitigate their effects? Well, people have come up with a whole suite of ideas to save wildlife from death by traffic. Some of these help a specific, charismatic species. Barn owls (*Tyto alba*), for example, have the habit of sitting on the hundred-meter markers that are such a common feature of highways in many European countries. From these perches they hunt for mice in the roadside verge by flying low across the road, during which they often connect sharply and fatally with traffic coming from the side. In a successful bid to reduce barn owl deaths, the Dutch barn owl conservation group has applied their own "hostile design" by installing rollers on the markers. An owl that tries to sit on one of these roadside signs will slide off so many times that eventually it will not even try anymore, opting instead for the new, much higher poles that the group installed near the markers. And these alternative perches are so high that the owls no longer fly into oncoming traffic.

It is a great idea, but it benefits just one of the many species that are killed on the road. A much broader applicability is had by installing fences. Roadside fences with a wide mesh width prevent large mammals from crossing the road. Smaller mesh widths also exclude smaller land-bound vertebrates, and better still (or in combination with taller wire mesh fences) is the completely impenetrable, unclimbable polyethylene fencing that is often applied along highways in places where roadkill, specifically of small mammals and amphibians, is very high.

The alternatives to barriers along the road are underpasses or overpasses, often in combination with fencing that can guide animals toward the wildlife bridge or culvert. The fanciest overpasses are the so-called ecoducts, which create a true landscape link between both sides of the road. On top of the overpass, soil and vegetation are planted to provide a natural-appearing corridor for wildlife to pass from one side of the road to the other. Some recently developed ecoducts are wonders of ecological engineering. There is one over the A4 motorway near Rotterdam that I pass every time I visit my mother. It is one hundred meters wide and consists of meadows, reedland, swamp, and a river, an entire wetland under which runs the five-lane highway between Rotterdam and Amsterdam. Combined with fencing, these ecoducts are the best thing we have today for preventing roadkill, at least for the nonflying animals.

But they are not good enough. Bill Laurance, a road ecologist at James Cook University in Cairns, Australia, says, "Many 'greening' measures, such as adding rope-bridges or underpasses to help species cross roads, bring relatively trivial benefits, akin to treating cancer with a Band-Aid." He is right. Even with the most advanced technologies of fencing, underpassing, and overpassing, roads still have a huge effect on the surrounding ecosystem. They kill anything that tries to cross the road away from the mitigation structures; they cause air, sound, and light pollution, drought in forest edges, and all the other negative impacts that the past decades of road ecology have uncovered. The problem is, despite all technological advances, our concept of "road" has remained unchanged since pre-Anthropocene times: it is a linear intervention on top of the landscape to connect two hubs of human settlement or other activity. In a world where those hubs are rapidly growing and multiplying, conventionally built roads are beginning to reach densities that are no longer tenable.

In my view, it is time to start doing two things. First, a radical rethinking of what a road is, is needed. Instead of considering it a priori as a structure placed *on* the surface, we need to begin building (and rebuilding) a type of road that is covered from beginning to end with one huge ecoduct that is as wide as the road is long—in other words, a tunnel. Only in that way can we seriously reduce the disastrous impact of roads on the living world. I realize this is little more than a pipe dream: when even simple mitigation measures are often already an uphill battle, we are nowhere near making this a reality; but still, it is worth a thought, even if it is conservation science fiction. In fact, something approaching what I have in mind, with added benefits, is being worked on by the solar panel company Studio Solarix. Their Butterfly Effect product is a lightweight net covered in thin, flexible, translucent solar panels that is suspended from an aluminum frame and envelops the road on all sides.

The second thing we need to do is, as members of a community, begin questioning the need for road-building projects in the first place. A group of people living in a neighborhood near Deventer, in the east of the Netherlands, did the latter. They approached the Taxon Foundation, the nonprofit that I founded in 2020 to help provide a scientific backing for community groups that study and protect nature and biodiversity in their immediate surroundings. In this case the group was concerned that the municipality wished to convert a centuries-old narrow sand path into a broader, paved, street-lighted road, which, they feared, would mean much more and much faster traffic and therefore much more roadkill for the wildlife in the area.

We devised a joint community science project. A team of biodiversity experts worked with a team of people from the neighborhood on a broad biodiversity inventory of both sides of the sand path. So, on a warm July Sunday, you might see the local general practitioner swinging a butterfly net alongside wasp expert Aglaia Bouma, a lawyer, dipping a water net in the ditch with entomologist Jan Wieringa, two schoolchildren sifting through leaf litter in a large white plastic tray, and a theater actor and me studying snails at a microscope installed in a makeshift lab in one of the sheds along the path. Together, in two weekends, our bioquest (the often used term "bioblitz" is really distasteful, reminiscent of Nazi Germany's indiscriminate bombing campaigns) unearthed no fewer than 376 species of invertebrates, many of which were rare, threatened, and red-listed. We wrote a solid scientific report and the community group presented this at the town hall to the assistant mayor, who was visibly impressed with the heavy-duty science and unable to deflect it as easily as the more familiar placards and slogans. The outcome was that the municipality changed its mind and ruled that the ancient sand path be left as it was.

Of course, I have no illusions that a single small community science project will ever be as powerful as the combined financial and lobbying weight of civil engineers, construction companies, and local politicians. But a hundred small community roadkill-monitoring projects might carry some weight, especially if some of those projects present their research data in the evocative form that Koese and his crew chose. The community science platforms are available, ready to be used by groups of citizens who take the initiative. And the more people who sat with a dying songbird in their hand by the side of the road or tried to usher a lizard off the tarmac, the more likely it is that such initiatives will be taken, and also the more likely it is that we will eventually learn to see roads from the environment's perspective and begin a radical rethinking of ways to transport people that do not necessarily decimate the biodiversity.

12

IT'S A TRAP!

Let us stay with roads for a little while longer. In March 2004, somewhere in Taiwan, the photographer Wilson Hsu captured a little tragedy in poignant detail. A dead barn swallow (*Hirundo rustica*) was lying face-down on the tarmac, killed by a passing car. A live, agitated, fellow barn swallow was by its side, in Hsu's words, "wishing to revive his dead relative. . . . He cried out, 'get up, get up.' Suddenly he approached closer, took hold of the other and tried to help it sit up. His dead relative was too heavy, but he kept trying time and again, while flapping his wings. In the end he used all the strength he could muster, but still received no response."

For a while, Hsu's series of photos trended worldwide. News media arranged them into a heartbreaking storyboard about a love tragedy. One social media post emblazoned the pictures with the all-caps text, "THEIR LOVE IS NO LESS REAL / THEIR GRIEF IS NO LESS PROFOUND / LOVE IS LOVE & LOSS IS LOSS." *Science Blogs* ran it under the title, "One of Life's Tiny Dramas Captured Forever on Film," with the author philosophizing, "It is heartbreaking to view them through the lens of human emotions, but I do wonder about the emotional lives of animals. . . . It is doubtful that birds think about (or obsess about) death as humans do, but does that mean they are less affected by death when it happens to a close companion? I guess this is one of the many great mysteries that we will never really know the answer to."

In his 2009 book, *The Duck Guy*, ornithologist Kees Moeliker claims to have an answer, and a soberingly prosaic one at that. To a bird expert, says Moeliker, the photos of the bird's behavior tell a different story. Dismissing all romanticizing, he writes, "The fluttering swallow-on-top was engaged in one of the best photographically documented cases of [animal] necrophilia."

Perhaps disturbingly, necrophilic sexual behavior is common in animals. Particularly, testosterone-laden males of polygamous species, including swallows, will try their luck on anything that remotely resembles a female in a willing pose. And a dead bird in mountable position on the road, rump sticking up invitingly, will do just fine (which, incidentally, is one of the reasons that epidemics in birds, like the avian flu, spread so rapidly). In the scientific literature, such copulations are often termed "Davian behavior" (after the lines in the limerick beginning "There once was a hermit named Dave / Who kept a dead whore in his cave," apparently). In fact, says Moeliker, who is himself the proud discoverer of the first case of homosexual necrophilia in the mallard duck (which won him the tongue-in-cheek 2003 Ig Nobel Prize for Biology) and tireless collector of data on "animals misbehaving," plenty of examples are known of animals having it on with road-killed conspecifics suffering varying degrees of being dead. Over the years, he has catalogued male pigeons, sparrows, three species of swallow (including the barn swallows of the previous page), toads, and frogs caught in the act of desperately trying to father offspring with both male and female corpses lying by the roadside.

One instance is particularly telling. In 2001, in the ordinarily parch-dry ornithological journal *Ardea*, says Moeliker, an article appeared by the Norwegian zoologist Svein Dale, who describes not only necrophilia in road-killed birds but also the first experimental approach to the phenomenon. On a highway in Greece, Dale writes, he observed a group of some two hundred sand martins (*Riparia riparia*) attending seven road-killed birds of that same species. To see if the group was really there because of their dead relatives, Dale picked up the seven corpses and moved them to a parking lot, and within two minutes the whole flock had moved there, too. Dale then kept ferrying around the dead birds (in all, six times), and each time the cortège settled exactly where the dead birds were. Tellingly, after each transfer, several of the live sand martins copulated with one or more of the dead ones.

Distracted as they are, many males thus helping themselves to dead (or nearly dead) partners on the road will become roadkill themselves, and the scientific papers describing Davian behavior often mention this knock-on effect. One sad case from Moeliker's collection was written up in a terse, detail-filled article in the 1985 volume of *British Birds* by a K. E. L. Simmons from Leicester, UK. The author witnessed how, on May 9, 1984, at 9:40 Greenwich Mean Time, a group of house sparrows (*Passer domesticus*) flew along a busy thoroughfare in the eastern part of the city. The straggler, a female, was hit by a car and landed limp on the road. She was, however, not dead, and over a few minutes slowly regained consciousness and assumed a somewhat stooped sitting position. At that moment, a male swooped down, hopped on to her, and copulated with her. He then dismounted, did a small courtship dance in front of her, and then, while she was still dazed, mated twice more until a passing car struck them both dead.

All this provides a new perspective on the roadkill we so exhaustively dealt with in the previous chapter: namely, that it is not always the accidental outcome of an animal being in the wrong place at the wrong time. Sometimes roads *attract* animals. This happens in the Davian examples that I just mentioned, where animals are sexually attracted to road-killed conspecifics. But there are more reasons why the lure of the road might be irresistible to many animals.

More likely even than attracting aroused mates, a corpse on the road will attract scavengers. Many foxes, ravens, buzzards, and vultures are killed on roads because they come to eat from the dead animals that litter the hard shoulders. Their corpses in turn attract more scavengers, and in this way a single roadkill event can become the nucleus for a snowballing of scavengers and scavengers-of-scavengers being killed, including, of course, the many insects that are attracted to the dead animals (flies, carrion beetles, wasps, and ants, for example).

Slow worms (*Anguis fragilis*, a kind of legless lizard) just love a warm surface *and* have poor eyesight—a deadly combination on the road, for they will often languidly stretch out on tarmac that is nicely warmed up in the sun, having no inkling of rapidly approaching vehicles. Or even not so rapidly, as ecologist Rob Bijlsma discovered. For nearly thirty years (Bijlsma is known for being an indefatigable recordkeeper of anything that strikes him as ecologically relevant), he counted slow worm deaths on two bicycle tracks in the Netherlands (with a combined length of 5.2 kilometers) and found that as bicycle traffic increased tenfold over that period, so did the dead slow worms, suggesting that it is simply a chance process: as there are more bicycles on the road, the chance of being hit by one before you have had enough of basking and move on, eventually approaches 100 percent.

And then there is the fact that many animals are attracted to roads for the same reason that humans use them, namely, they are pleasantly flat and smooth places for traveling along. Firefly larvae, for example, often use roads to disperse to suitable spots for them to pupate and are then killed by cars, despite carrying proper lighting.

In colder climes, some animals are also attracted to the salt that is used in winter to de-ice the roads. Many mammals, especially herbivores, live under constant threat of sodium deficiency, and they will often congregate at natural salt licks, places where salt and other minerals come to the surface. Unsurprisingly, the salt spray that sticks to roadside verges along highways in winter is similarly irresistible to them, again leading to roadkill deaths.

CARNAGE IN A BOTTLE

All these examples that I have just mentioned, of animals following some deep-seated natural urge into a dangerous human-made place (in this case, a road), are instances of what biologists call an evolutionary trap—evolutionary because evolution has provided the animal in question with the strong motivation to look for a mate or a meal, or a flat surface. And whereas under natural circumstances, such behavior is advantageous, the urge in this case lures the animal into a deadly trap.

Sometimes an evolutionary trap takes quite a literal form, coming complete with bait and a trapdoor of sorts. To see what I mean, next time you spot a bottle or a soda can lying in a tilted position on the ground, pick it up and shake out the contents in a tray. There is a good chance that on closer inspection, the soup inside is a slurry made up of hundreds of animal fragments.

What probably happened is that a small mouse or shrew looking for a burrow climbed into the can or bottle. Or perhaps the yeasty smell of leftover beer or the sweet, fruity scent from the soft drink attracted some insects (think flies, ants, beetles, or wasps). Then the sun heated up the container and the furious pitter-patter of tiny feet endlessly slipping on the smooth inside was heard for a while until the prisoners, unable to reach the neck of the bottle or the tear strip of the can, perished one by one. The dead animals inside began to rot and emit the odor of death, and this in turn attracted carrion-feeding insects such as bluebottle flies and carrion beetles. They, too, died, which only enhanced the intensity of the scent and attracted more and more insects into the evolutionary trap until it was filled with hundreds—nay, thousands—of dead and decaying animals. In this way a single empty bottle, especially when lying with its neck pointing up, can act as a vacuum cleaner, a pit of death that decimates populations of several insect species in the local area.

We know very little of this kind of evolutionary trap. For the past few decades, the occasional zoologist has looked at dead mice and shrews trapped in discarded litter. In Italy in 1997, for example, Paolo Debernardi of the Research Centre for Applied Ecology in Torino, picked up 189 bottles and six cans and inside found the remains of 904 small mammals. Some bottles were true mass graves, containing up to thirty-two corpses assignable to four different species. In August 2014, a field team of Spanish biologists chanced upon a two-liter soda bottle in a beech forest that was found to contain "a brownish broth resulting from the mixture of hair and bones of small mammals." After bringing the bottle to the lab, the researchers filtered and cleaned all the animal bones and teeth inside and studied them under the microscope. They concluded that the bottle contained the cadavers of no fewer than fifty-four small mammals belonging to three species. Sure, it was a large bottle, but this is still the world record.

But the much more numerous insects and other invertebrates trapped and killed inside have almost not been seriously studied yet. There are just one or two cases in which an attempt has been made, and only in very recent years. In the spring of 2019, Federico Romiti of the University of Rome and his students collected 172 bottles, tanks, jars, and cans on a beach in southern Italy. In them, they found the remains of 2,811 beetles belonging to thirty-eight different species. And that was just the larger beetles. There were probably similar numbers of dead snails, millipedes, bluebottles, and wasps inside, not to mention the even tinier creatures.

Almost nobody has even tried to put numbers to the death toll that disposable bottles impose on such smaller animals.

The problem of animals trapped in Anthropocene litter is so pervasive and global in scale that it cannot really be properly assessed by small studies by one or a few university researchers. Just as with roadkill, we need to begin harnessing the power of social media and community science. Tabulating the animals evolutionarily trapped inside bottles and cans can be perfect urban community science projects, to be conducted either through the combined efforts of the global community of keen observers and photographers or by single driven individuals.

One such driven individual is Mr. Graham Moates, who lives near Norwich, UK. He is the type of wonderful community scientist-naturalist we keep coming across in this book. Although professionally a chemist working for a food science institute, he is often outdoors, taking photographs of animals and plants. He documents the moths and other insect biodiversity of his garden for the UK-based citizen science platform iRecord, volunteers for the Norwich Bat Group and local nature reserves, and is an active organizer of litter cleanups in his hometown.

During those cleanups, he occasionally noticed dead mice and voles trapped in the bottles he picked up with his litter tongs. So he decided to begin a small scientific study of his own. Over the course of several litter-picking seasons in 2016 and 2017 in and around his hometown, he collected nearly 2,200 uncapped and uncrushed cans and bottles. For each, before he stuck it into his litter collecting bag, he jotted down the dimensions and the material, and poured out the contents into a tray, which he then preserved in a ziplock bag and took home. There he painstakingly sifted through the contents of each bottle to pick out any mammal skulls and jaws, which he later identified using the identification keys in his booklet *The Analysis of Owl Pellets*, a publication by the UK Mammal Society.

Writing up his results for his 2018 article in the *Journal of Litter and Environmental Quality*, he calculated that 5 percent of the bottles and cans contained remnants of a total of 230 dead mammals. "I never thought it could be this many before I began the research," he told the UK newspaper the *Daily Mail*. "It is very upsetting to think of how many small creatures die in this way because of human carelessness."

The nice thing about Moates's community science project is not just that he used rubbish cleanup campaigns as a way to do important research but also that he went quite far in teasing information out of the data he had collected. For example, he also measured the widths of the bottlenecks and, comparing these with information from the 1960s, found that these days, bottle openings are smaller and more uniform than in the past—probably the result of industrial standardization. For the mammalian deaths this means that today, mostly the smaller shrews would enter the evolutionary traps, rather than the somewhat larger mice and voles that would have been commoner victims in the past.

Another way in which Moates gave an interesting new spin to his data was by applying his professional knowledge as a food technologist to them. Being aware that the UK produces a staggering ten trillion soda and beer cans a year, that some three trillion of those are not recycled, that around 350 million cans are picked up in litter cleanups, and that about 6 percent of the littered cans are overlooked, he did some number crunching and came up with a figure of 27.4 million containers annually available for small mammals to crawl into. Since his research showed that this happens to about 5 percent of the containers, and that the occupied containers contained on average slightly more than two dead animals, he calculated that yearly, some three million mammals in Britain die this way.

That figure was picked up by the popular UK presenter and conservationist Chris Packham, who is also an ambassador for the Keep Britain Tidy campaign (the blunt slogan of which is "Don't Be a Tosser!"), which led to Moates's one-man community science project to be in the headlines for several days in March 2018, hopefully triggering others to do similar studies in their local litter-picking campaigns.

The opposite of Moates's one-man project was a study carried out by Krzysztof Kolenda and colleagues at the University of Wrocław, Poland. Although the team members were all professional academics, the data they used consisted entirely of pictures on Instagram, Facebook, YouTube, Twitter, and other social media posted by ordinary people who had come across animals trapped in discarded bottles and cans. For five months in 2019, the team trawled the social media in nine languages, looking for words and hashtags like #trappedanimal and "bottle" or "can." They found nearly five hundred posts from six continents concerning more than a thousand trapped animals, ranging from an Australian possum stuck inside a Nutella jar, to a collection of rare endemic insects, scorpions, and lizards dead inside a Corona beer bottle abandoned in the Chilean Atacama Desert, to seventy dung beetles found in a Polish beer can.

Although the reports came primarily from urbanized areas, they were from fifty-one countries and concerned all manner of containers and animals, an indication that this is truly a global problem. Still, all these little bottlenecks pale by comparison with an evolutionary trap that has a global gape width of thousands of square kilometers—and counting.

TRICK OF THE LIGHT

My friend Bernhard van Vondel is one of the world's foremost experts on water beetles. A construction engineer by profession, he picked up his hobby of insect collecting by accident while on holiday in Spain many years ago. Today, his is one of the world's most valuable water beetle collections, and his expertise is nearly unparalleled. A balding, bearded, bulky bear of a man, it is endearing to watch him lovingly dissect and sketch two-millimeter-long beetles. And although he mostly busies himself with classification and taxonomy, in 1987 he made a small water beetle–related discovery that turned out to have far-reaching consequences for evolutionary traps and their impact on technology for the energy transition.

That year, while camping with his family near the Loire river in central France, Van Vondel noticed how his old Renault 4 GTL could be helpful in a completely unexpected way besides transporting his family and camping gear. For onto the roof of the little red car descended, from around 4 p.m. until 8 p.m. each day, a great variety of water beetles. So, instead of lugging his heavy-duty water net to ponds and streams to collect his water beetles, all Van Vondel needed to do was to sit in the camp site with his pooter (a contraption to suck up tiny insects into a vial) next to his car, cold beer at hand, and simply pick the beetles off the paintwork as they settled there. Over the course of two days, he collected 123 water beetles this way, belonging to twelve different species.

In 2004, the little half-page article Van Vondel wrote about his lazy approach to insect collecting was one of the triggers for the Hungarian physicist Gábor Horváth, director of the Laboratory of Environmental Optics in Budapest, to begin a study on "the attraction of water insects to car paintwork."

Horváth is a prolific researcher who has devoted his career to the way animals use the polarization of light to find their habitat. You see, light can be thought of as a wave. Like the waves we can produce in a rope by fixing one end and holding the other and then moving it up and down, the direction of a light wave lies in a certain plane. With the rope, if our hand moves up and down, the wave will also be vertical; if our hand shakes side to side, the wave will be horizontal. It is the same with light waves. In the light that comes from the sun, all rays have different directions of oscillation, from horizontal to vertical and any diagonal in between. But a polarization filter, like the one in polarizing glasses, can cut out waves in all but one direction. Many flat, shiny surfaces also work as a polarization filter and turn the ragbag of directions of oscillation into predominantly horizontal ones.

Humans are not very good at detecting the direction of light polarization, but many insects have perfected the skill and use it for navigation. And that is where Van Vondel's Renault 4 comes in. Since horizontal surfaces of ponds and streams polarize the light by converting up to 70 percent of the sunlight into horizontally polarized rays of light, water insects (which fly around looking for places to settle or to lay their eggs) home in on any large patches of horizontally polarized light that their eyes pick up. For millions of years, those patches would have been reliable cues for bodies of water. But today, humans produce surfaces much flatter and shinier than even a still pond—in other words, perfect evolutionary traps that trick water insects into landing on them.

Over his thirty-five-year career, Horváth has been revealing these mirage-like evolutionary traps one after another. Back in the mid-1990s, he discovered that oil lakes in the Kuwaiti desert (formed during the First Gulf War when the Iraqi troops blew up hundreds of oil wells) have even stronger polarizing effects than water and attract enormous numbers of dragonflies, mayflies, and other water insects that try to lay their eggs in the lakes and immediately are enveloped by the sticky goo and die. He and his students then went on to discover many other similar death traps: asphalt roads, the glass windows of buildings, and shiny polished gravestones in cemeteries all turned out to produce even more strongly horizontally polarized light than water surfaces and therefore irresistible to water insects. The aquatic animals congregate on these surfaces, which often have been heated up by the sun, and die in huge numbers; or they lay (to waste) their entire egg supply on the inappropriate surfaces. (Incidentally, the discovery that dragonflies lay eggs on polished gravestones won Horváth the 2016 Ig Nobel Prize for Physics—evolutionary traps have always been a good field of study for winning Ig Nobel Prizes.)

In 2004, prompted by Van Vondel's discovery, Horváth also tried his hand at (variously colored) car roofs. Together with György Kriska and other colleagues and students, he first placed shiny plastic sheets in four different colors (black, red, yellow, and white) on a lawn near a swampy area and picked off all the aquatic insects that alighted on them one summer evening. They found that of the 1,229 insects (belonging to thirty-seven species), nearly 90 percent landed on the black and red sheets. Next, they called up the director of Suzuki Top in Budapest and somehow persuaded him to lend them Suzuki Swifts in each of those four colors. The researchers then parked the borrowed cars in the sun and measured the polarization of the light that reflected off them. The red and the black Suzukis turned out to have strong polarization properties, but the white and yellow cars hardly any at all. No wonder it is mostly the owners of red and black cars that find their roofs covered in floundering water beetles, cooked water bugs, and doomed mayfly eggs!

Given the numbers of parked cars in the world, Horváth and his team say, even when discounting the yellow and white ones, there is a gigantic "car pool" for insects to be evolutionarily lured into. But Horváth's most recent work shows that an even greater danger lies in solar panels. These saviors of the energy transition are being installed at an increasing pace all over the world, and they, too, are strong water insect attractors. With the surface covered by these so-called photovoltaics growing nearly exponentially, their role in insect declines should also be becoming a concern. Although most people seem unaware of or unconcerned by their insect-killing properties, Horváth has been pioneering solutions to reduce the attractiveness of the panels. He discovered that by fragmenting the appearance of a panel into multiple smaller squares by arranging a few white strips across them, they lost nearly all their trapping effects. And in a follow-up study, a team he was part of found that if solar panels were coated with a thin puckered film, they also no longer polarized the light. Recent research by Horváth's colleague Bruce Robertson shows that it is not only aquatic insects that are misled by solar panels. Water birds similarly collide with solar panel installations because they, too, use polarized light to home in on water bodies.

NAVIGATING THE MINEFIELD

Life in our urban environment is like navigating a minefield of evolutionary traps. In this chapter, we have had a look at roads with all their deadly attractions, at literal traps disguised as yummy-smelling bottles and cans, and at the misleading glare of the fake lakes of the Anthropocene. And there are many more. Some are actually intended to kill animals, like the pheromone traps that mimic the irresistible scent of a willing mate and thus eradicate pest insects. Most are accidental but nearly impossible to steer clear of. Artificial light at night misleads migrating birds and insects, imprinted as they are by eons of evolution to keep the brightest hemisphere (which, even at night, was always the sky) above them. And then there are the innocent-looking shortcuts that make birds (who still do not understand glass) fatally slam into windowpanes, or the comfortable-looking pylons that electrocute perching birds. We ourselves also fall into an evolutionary trap when we gorge ourselves on illness-inducing fatty, salty, and sugary foods, prepared by thousands of years of hardship to gobble up these energy-rich resources whenever we come across them.

Even when we try to help urban animals, we may be leading them unwittingly into an evolutionary trap. In Illinois, for example, conservationists fixed nest boxes to trees to aid the local population of wood ducks (*Aix sponsa*). However, these ducks normally nest in well-hidden places to prevent neighboring females from coming around and dumping eggs, which are then incubated by the foster mother (so-called "intraspecific nest parasitism," basically what cuckoos and cowbirds do but within the same species). With the nest boxes so easily visible, females were dumping their eggs left, right, and center, leading to some nests with up to fifty neighbors' eggs to be incubated by a single unlucky duck, and most of these oversized clutches failed altogether. By providing the wood ducks with a lot of spacious, easily accessible artificial nesting sites, the conservationists had actually set up a perfect evolutionary trap for the species and nearly driven the population to extinction.

The problem with evolutionary traps is that they turn an animal's environment on its head in an irreparable way. For millions of years a certain cue was a reliable signal for an animal that some benefit—a willing mate, a suitable habitat, a favorite kind of food—was to be had there. But because humans, mostly accidentally, produce a replicate of this signal, the animal's "information" has suddenly become outdated. Since time immemorial, water beetles would reliably home in on horizontally polarized light, a cue so dependable that the behavior has become hardwired into the insects' nervous system. But all of a sudden—sudden in evolutionary terms—the insects can no longer trust their own eyes.

And because these cues have been so reliable for so long, and because that reliability has been hardwired in the animals' genes, behavior, and physiology, their sudden unreliability cannot easily be reverted by evolution. A few chapters later I will show examples of urban challenges that animals and plants *can* adapt to through rapid evolution, but these adaptations usually require relatively simple tweaks in their genetic makeup: a change in color, or the size of their seeds, or the time of the year they mate. But evolutionary traps are sprung so relentlessly and permanently that organisms often simply cannot evolve their way out of them.

That is all the more reason for urban community scientists to closely monitor the havoc these evolutionary traps wreak: by counting dead animals inside littered cans and bottles, as Graham Moates did, but also, for example, by observing how many water insects die on solar panels or car roofs. Or by monitoring bird deaths by window strikes, as is done in community science projects such as New York City's Project Safe Flight and Toronto's Fatal Light Awareness Program. Or indeed by finding an evolutionary trap of your own and just starting to investigate it. This is important because we need the data to gauge the death toll that all these traps take, but also because often it is not that hard to disarm an evolutionary trap, if only people were aware of the existence of such traps and the scale of the problem.

Let us consider the following scenario. Many homeowners make changes to their home and garden to do their bit for the environment.

They install solar panels on the roof to help in the energy transition, and they excavate a pond in the garden to help amphibians and aquatic insects. You, a budding urban naturalist, now know that many of the insects that develop in and emerge from that pond will probably, on their maiden flight, land and die on the solar panels on the roof of the same house—a sobering message. But you also know that the impact can be easily and cheaply reduced to nearly zero by the crosswise application of some white tape to the panels (something no solar panel company has yet implemented). So why not start a neighborhood initiative? Find some neighbors who have both ponds and solar panels, apply strips of white masking tape to half the panels, and spend a comfortable evening (beers and snacks are allowed) sitting on their roof collecting and identifying water insects as they descend (or not) on the panels. If you manage to replicate Horváth's results, you have solid scientific leverage to convince all your neighbors to apply this simple mitigation measure to their solar panels, and to help make the neighborhood a better place. That is what community science is all about.

13

ANIMAL ARCHITECTS OF THE ANTHROPOCENE

In the previous chapter, we saw how traffic, discarded bottles and cans, and smooth surfaces that optically mimic ponds and lakes are just a few of the many human inventions that indiscriminately and unexpectedly kill wildlife in urbanized areas. With top predators largely absent from many cities, these unintentional evolutionary traps take their place as an important source of animal mortality. But some human intrusions, though potentially deadly, can have a much more complex and nuanced impact on the urban ecosystem than you would think at first glance. One such impact is—literally—showcased in a museum in my hometown.

It is not a large museum. In fact, it is the smallest museum you have ever seen, housed in a disused bridgekeeper's house of less than ten square meters attached to the side of an old steel drawbridge in the city center. "Exhibition" is announced in quasi-Art Deco lettering on the outside. As befits a bridgekeeper's house, it has windows all around, and behind those windows is a genuine but very unlikely museum display of urban "natural history" finds: a waterfowl nest with various plastic objects in it, a dead duck garrotted by a plastic bottle ring, plastic shower gel containers from the 1970s, an alcohol-filled jar in which floats a perch stuck in a rubber glove, various sex toys covered in algae and freshwater clams, and more.

The bridgekeeper's house is not just a museum, it is also the headquarters of De Grachtwacht, which means "Canal Watch" and is a humorous reference to *Nachtwacht*, a famous painting by Rembrandt, who was born just down the street. The Canal Watch's main activity is to organize weekly "Plastic Spotter" cleanups of the city canals, the spoils of which are highlighted in this museum. Each Sunday an armada of green canoes led by the biologist couple Liselotte Rambonnet and Auke-Florian Hiemstra and their team of volunteers and community scientists paddles a set route through the many waterways in the old city center. Each two-person canoe is armed with a large bucket lined with a heavy-duty garbage bag and a grabber. One volunteer paddles, the other grabs and bags each artificial object they spot floating in the water.

But after the cleanup, as soon as the six-canoe team haul their buckets up on the quayside and turn their backs, new artificial debris floats by behind them. If cleaning up alone were the aim of the Canal Watch, the participants would soon run out of steam because of the sheer futility of their task. The twice-weekly open market, the many summer festivals, and the frequent westerly storms dump much more plastic into the canals than even a *daily* Plastic Spotter operation could clean up.

That is why the Canal Watch's main aim is not just to clean up but to create awareness and, even more important, to collect data on the plastic pollution problem. After every cleanup they do not simply bring their catch to the nearest recycling plant. Instead, Rambonnet, Hiemstra, and their volunteers carefully sort, catalogue, and count all the items they have pulled out of the water. Plastic bottles, plastic straws, toys, bits of polystyrene foam, glass bottles, aluminum cans, takeout coffee cups, candy wrappers, plastic shopping bags, deflated balloons, sexy underwear, cigarette butts, laminated menu cards from restaurants, and, during the COVID-19 pandemic, face masks and rubber gloves—all are stored separately, counted, photographed, and tabulated.

The trophies from all this painstaking and sometimes downright dirty work are displayed in the museum's display cases. But we also find a more indirect output there: prints of scientific journal articles authored by the Canal Watch team, papers with titles like "The Effects of Covid-19 Litter on Animal Life," "Birds Using Artificial Plants as Nesting Material," and "Plastic Hotspot Mapping in Urban Water Systems." Like Graham Moates of the previous chapter, Rambonnet and Hiemstra and their team are using plastic cleanups as a source of scientific information on the environmental impact of human litter. And they are finding that the impact of plastic is more complex than it seems at first sight—especially when you start looking, as Hiemstra does for his doctoral studies at Leiden University, at animals that build.

PLASTIC FANTASTIC

By international standards, with my 1.86 meters I am fairly tall, but in the Netherlands I am of average height at best. So even I feel dwarfed standing next to Hiemstra, who is a burly Frysian at least fifteen centimeters my superior—thirty if you count his natural Afro of frizzy reddish hair. As he welcomes me in the doorway of the tiny bridgekeeper's house, his presence looms even larger, but as soon as he takes my coat and starts showing me some of the objects in his library and display collection, his demeanor morphs to match their delicacy and his fondness for them. A mini-diorama with a stuffed bittern, of which he gently caresses the chest feathers; a nest of the extinct passenger pigeon; and plates from a volume of *De Nederlandsche Vogelen* by Nozeman and Sepp, the earliest book on the birds of the Netherlands.

Hiemstra is a curious hybrid of traditional natural historian and modern ecologist of the Anthropocene. His love for old books, natural history curiosities, and museum displays fits his traditional half, whereas his modern half shows in the snappy children's book on city nature that he wrote, his TV appearances in kids' programs, and his work for the Canal Watch. And he navigates effortlessly between these two personas. Gently leafing through the Nozeman and Sepp book, he explains to me how two hundred years ago this outsize book was for a while the most expensive book in the Low Countries. It showcased nearly two hundred species of birds, in full color and, where possible, life size (hence the large format).

Hiemstra shows me the title page and points out that the subtitle to *De Nederlandsche Vogelen* (The Birds of the Netherlands) is *Volgens Hunne Huishouding, Aert, en Eigenschappen Beschreeven*—Described by Their Housekeeping, Character, and Peculiarities. Then he turns to page 63 and shows me the plate of the nest of a coot, *Fulica atra*. Nest building is one aspect of the "housekeeping" of the subtitle, it appears. In the eighteenth century the coot (a compact velvety sooty bird with a bright white blaze on its forehead) was still a bird of remote swamps and lakes, where it would breed on a nest made of reeds and roots.

Nowadays, says Hiemstra, coot nests look very different. From a shelf, he pulls down a stuffed coot nailed to a plank that he once bought in a flea market. "This one dates from 1950 and I like to think it is the last

Dutch coot that grew up in a nest like the one in the book." For today's coots always come into the world in nests partially or entirely made from artificial materials. In the decades after World War II, plastics rapidly became widely available and penetrated the environment at an exponential rate. Today, by weight, we have twice as much plastic sitting around as there are animals on Earth, and the amount of plastic produced yearly (which now stands at 500 million metric tons) continues to grow. At a certain moment, the coot's environment became so cluttered with pieces of plastic that the birds began incorporating them into their nests. This was helped by the fact that, at some point during the twentieth century, coots developed an urban streak. Whereas they were countryside birds in the past, they now are the most common species of waterfowl in many European cities.

For his PhD research, Hiemstra studies the ways in which coots incorporate plastic waste into their nests. On his laptop, he flips through photo after photo of urban coots nesting in constructions assembled from a wild variety of human-made objects. One nest from a canal in Amsterdam contained more than a thousand plastic items, including coffee cups, ladles, fireworks, bits of string and ribbon, garbage bags, earplugs (ironic, says Hiemstra, given how loud coots can be), and lots and lots of candy wrappers.

To really get to grips with the coots' nest-building strategies, Hiemstra uses his Canal Watch canoe to collect nests after the nesting season. He then carefully dissects them, keeping complete logs of all the objects he encounters. As it turns out, even nests that at first sight seem completely natural, made from twigs and leaves, on dissection often turn out to contain dark, algae-covered pieces of plastic in between the leaves and translucent nylon fishing line woven in with the twigs and branches. Often the building materials neatly reflect the predominant human subculture in the vicinity. In proper residential neighborhoods, he finds nests made from high-end supermarket shopping bags, fancy ice cream treat wrappers, and VIP festival wristbands. In inner-city Amsterdam, nest materials might include condoms, sunglasses, and bicycle tires, and many items related to recreational drug use (ziplock cocaine bags, plastic cannabis joint tubes, the occasional disposable syringe wrapper). During the COVID-19 pandemic, rubber gloves, self-test swabs, and face masks—up to twenty-seven of these in a single nest—were all the rage. One coot (perhaps in an attempt to build a car, Hiemstra quips) had windscreen wipers and a seat belt in its nest. "Every time I dissect a nest it's still a shock," he says.

When he lectures to students, he often shows images of some of the more extremely "polluted" nests, gaudy mosaics of the debris of human wastefulness and decadence, with plastic bags, drinking straws, and chunks of polystyrene foam sticking out on all sides in a profusion of colors. Whenever he asks the students what these images elicit in them, they invariably reply with "sadness," "disgust," or some similar negative emotion. And although initially this was also Hiemstra's overriding feeling when he began to study the coots' artificial nests, his view soon began to shift.

EARLY ADOPTERS

Although some plastic objects appear to have been piled on and around the nest in haphazard fashion, others are incorporated in a more "creative" manner—for example, as replacements for natural building materials. Some coots use plastic artificial plants in lieu of real ones to build up the nest, and one coot at the Uniper power station in Leiden had pleated fifty-three plastic drinking straws around the nest cup, pretty much exactly in the manner that those eighteenth-century coots in the Nozeman and Sepp book did with stalks of reeds. Several nests had some long, flat plastic packing straps deliberately wound around the nest several times. And when he took apart the bowl-like cores of nests, Hiemstra would often find layers and layers of plastic sheets tightly stacked, producing a sturdy, layered mattress.

In the heart of Amsterdam, in a canal along Rokin street, Hiemstra and his student Atze van der Goot found a site where consecutive generations of such layered plastic nest bowls had been built one on top of another. Judging by the dates printed on the food wrappers they found, starting at the bottom with a Mars bar wrapper advertising the FIFA World Cup of 1994 and continuing upward with over thirty candy wrappers, soft drink packages, and soy sauce bags with more and more recent sell-by dates, urban coots had been doing this year in, year out for nearly three decades. Excavating these plastic sediments, Hiemstra says, made them feel almost like paleontologists of the Anthropocene.

Perhaps, Hiemstra began to think, coots do not build their nests from plastic out of sad desperation but because it is an improvement over their old nest-building techniques. They might be appreciating in these artificial materials the same properties that humans have created them for: strength, durability, and water resistance. A coot that shifts away from natural materials may enjoy a nest that is better at withstanding the wear and tear from wind and waves, does not decay and disintegrate, and might even maintain its body heat better. Perhaps parasites do not accumulate in a plastic nest, and its sturdiness may allow a pair of coots to use it repeatedly and raise more chicks. In other words, a modernistic coot with a penchant for plastic may be better off than an old-fashioned one that stubbornly sticks with natural materials.

But is this really so? Hiemstra is not sure yet, but he is dying to find out. With a physics student, he has been studying the physical properties of coot nest bowls with and without plastic, and preliminary results indicate that nest bowls indeed cool off more slowly when plastic has been incorporated into them. And the urban paleontology project at the Amsterdam Rokin nesting site seems to show that plastic nests can be reused over many years, something that coots never did when they were still using reeds and leaves (those nests would require constant maintenance and never survive more than one nesting season). Tellingly, when he conducted a project in which thousands of people from all over the Netherlands sent in pictures of coot nests, they discovered that some coots consistently build with plastic, even when plenty of natural materials are available. And, in contrast to these early adopter coots, there are also laggards: coots that stick with twigs and leaves in the presence of an abundance of plastic building materials. This is another indication that coots do not randomly pick up plastics by accident but that a change in their nest-building behavior is really taking place, with some coots making the change earlier and more drastically than others.

The coot discoveries are emblematic for many other animals that construct dwellings in an Anthropocene world. Many bird species worldwide do as coots do and incorporate artificial materials in their nests. Sometimes they go all-out, as when urban carrion and jungle crows in Japan

(*Corvus corone* and *C. macrorhynchos*) build nests entirely from colorful clothes hangers that they steal from apartment building washing lines. Hiemstra himself is also studying the mole (*Talpa europaea*), an underground mammal that builds nest chambers in the soil; as he discovers when he digs them up, the chambers are built more and more from pieces of plastic, especially in city parks. The same goes for those moth-like insects called caddisflies: their larvae live in freshwater in tubes they build from sand, shells, and sticks, and, as Hiemstra discovered, nowadays also from microplastics.

The ways in which animals use artificial materials suggest that more is going on than simply a dearth of conventional building materials. In Mexico City, the ornithologist Monserrat Suárez-Rodriguez discovered that house sparrows (*Passer domesticus*) and house finches (*Carpodacus mexicanus*) stick discarded cigarette butts in their nests. The more butts in a nest, the less the chicks in the nest suffer from blood-sucking mites. As it turns out, the leftover nicotine in the butts acts as an insecticide. In Spain, researcher Fabrizio Sergio found that black kites (*Milvus migrans*) decorate their nests with white ribbons of plastic that they tear from discarded shopping bags. Sergio and his team did experiments that not only showed that the birds have a preference for white plastic over other white materials and over plastic of other colors but also that the nest decoration is an important but mysterious signal in the battle over territories. Owners of nests with more white plastic strips somehow were better at chasing off rivals, even when those items had been placed on the nest by the researchers rather than by the birds themselves.

The Polish researchers Zuzanna Jagiello, Łukasz Dylewski, and Marta Szulkin discovered that on beaches worldwide, at least ten of the sixteen different species of Coenobitidae (land hermit crab) today use plastic, glass, and metal bottle caps and other cup-shaped trash in lieu of empty snail shells. They found this out with a method they call "iEcology": scouring the internet for funky snapshots taken by holidaymakers, community scientists, and whoever else would photograph a hermit crab in an unusual home—a bit like the study of animals killed in bottles of the previous chapter. They found 386 such photos, and they speculate that this trend may not be due solely to a scarcity of suitable shells. Instead, the plastic homes might be lighter than snail shells of the same size, and so easier to carry around. They may also be more colorful or more interestingly shaped. As hermit crab females prefer males with more outlandish homes, this could give the plastic junkies an edge in the mating game. If true, then well-meant campaigns such as the one by wildlife photographer Shawn Miller, who coaxes hermit crabs to swap their plastic homes for proper shells (which earned him tens of millions of views on TikTok), may not actually be helping the animals.

And Hiemstra himself helped reveal what may well be the irony of ironies in bird nest design: a magpie (*Pica pica*) nest made from antibird spikes! Eager for the public to send him sightings of weird nests, in 2021, he received an email from a patient in the University Hospital in Antwerp who, from his hospital bed, could see a magpie nest in the hospital's courtyard for which the nesting pair had apparently torn off a large part (148 strips, totaling almost fifty meters, as it turned out when Hiemstra carefully studied the nest) of the antibird strips that the hospital had glued to the edge of the building's roof. Magpies create dome-shaped nests, and in this case the strips had been strategically placed in the roof of the nest, with the spikes pointing up, to make it difficult for nest-raiding crows to enter the nest and steal the eggs or the chicks. In this case the very item of hostile design intended to deter birds had actually been appropriated by the nest-building birds to deter *other* birds!

Hiemstra, of course, would be the last one to claim that plastic in the environment is a good thing, and I would not claim that either. Plastics clog the digestive system of seabirds; they release chemical compounds that mimic hormones and mess with the physiology of mollusks and fish; they break down into microplastics that insinuate themselves into animals, plants, water, and air, even into the cells in our own bodies. Even if they have benefits as building materials for some birds' nests, they doubtlessly also have negative effects: chicks and parent birds can get entangled, and a plastic-lined nest bowl may fill up with rainwater and cause the nestlings to drown.

The important lesson here is that good and bad or natural and unnatural do not exist in nature, only in human minds. To the urban ecosystem, the omnipresence of plastic is not good or bad; it just *is*. As with all ecological features, for better or for worse, the organisms in the system will tend to adapt to its presence. And that makes for a more complicated relationship with pollution. Toward the end of this book we will meet Polish artist Diana Lelonek, who elevates overgrown human-made objects to botanical displays in the botanic garden of Poznań. She says, "One scientist found a very rare kind of moss on a piece of Styrofoam that I found. This is also a question, if you find a piece of Styrofoam in the forest, covered by this protected species, from the Red List, what should you do? Leave the trash in the forest, or throw it away? This shows that traditional thinking about nature doesn't apply here. In the botanical atlas it says this kind of moss usually grows on stones and rocks next to a river. So maybe we should extend this classification: moss also likes to grow on Styrofoam that's next to a river."

In other words, what if our waste becomes incorporated as an essential part of the natural environment? Its ubiquity and abundance have already caused animals and plants to begin adapting to its presence, to make it part of their ecology and way of life. As this is happening, do we need a rethink about when it is still pollution and when it becomes so inexorably appropriated and adopted by the flora and fauna around us that it becomes integral to the ecosystem? It may seem obvious that cleanups of plastic and other trash in the environment is a way of righting a wrong, but it may already be too late. Perhaps coots, kites, and mosses have already become dependent on our refuse and removing it from their environment could harm them in some way. I am not saying that this is the case, and probably the benefits of cleanups still outweigh their disadvantages, but it is important to realize that our impact on the environment may be irreversible, and that attempts to undo the damage do damage themselves.

In a living, evolving world, the environment may immediately and irrevocably respond to our actions, and soon things may have gone so far that it becomes impossible to reverse the situation to its original state. This notion is particularly relevant to our introduction into the environment not of inert materials but of living organisms. Such exotic intrusions are taken up in the next chapter.

LENS MADE IN SINGAPORE
2.8/85

14

THE ACCIDENTAL ECOSYSTEM

The imposing modernistic Belgrade Museum of Contemporary Art is the only building of significance among the vast expanse of lawns, paths, and shrubberies in the north of the modern township of Novi Beograd. Through the large windows of the museum's exhibition halls we see the midday sun beating down on the park outside and groups of schoolchildren, families with strollers, and lone joggers and cyclists traveling to and from the brutalist residential areas in the distance. But inside the museum it is pleasantly cool and quiet. We are visiting the exhibition *City: A Place of Identity*, and I am watching a video installation from 1977 by the conceptual artist Neša Paripović, a 16mm film in which the artist follows an imaginary line drawn straight through summertime Belgrade.

Dressed in a stylish maroon velvet suit with bell bottom pants, he hastily traverses the city. We see him crossing playgrounds, scaling walls, pacing across roofs, and scooting down steep slopes. I feel an odd kind of affinity with the artist because I spent the past week behaving in a very similar way while sampling the urban biodiversity of Belgrade. Shoppers on Kneza Mihaila saw me running after fast *Amara* and *Harpalus* ground beetles as they raced across the pavement in the hot midday sun. In Studentski Park, I was observed darting sideways and whipping out a butterfly net to catch a passing hornet, and I also clambered up the ancient walls of the Kalemegdan fortress looking for scorpions and slugs. In fact, as this thought crosses my mind, I notice a dot moving across the projection screen that is clearly not part of the video. As I move closer, I see that it is a large shield bug, one that has apparently flown in through the open transom windows. I wait for the guard in the corner of the exhibition hall to look the other way, then quickly move toward the projection screen and pick it up. Holding it by the pointy sides of its pronotum, I bring it closer to my eyes and confirm what I already suspected: it is the brown marmorated stinkbug, *Halyomorpha halys*.

Hailing from East Asia, where it is a very common insect, this large bug (flat, about one and a half centimeters long, with an attractive mottled pattern of many subtly different shades of brown) has been spreading in Serbia and other European countries, after first having colonized North America. It can live on the fruits and seeds of a broad spectrum of plants, both wild and cultivated, and has a habit of overwintering in sheltered places such as sheds, barns, and houses. As a result, it has been hiding in agricultural produce and passengers' luggage and accidentally hitchhiking by plane to all corners of the world. In Serbia, it was first noticed in October 2015, when a sighting was posted on a Facebook group for naturalists. Since then, its populations have multiplied year by year. Today, says philosopher Andrija Filipović of the Faculty of Media and Communications at Singidunum University in Belgrade, they are one of the most familiar insects to the people of Belgrade, not least thanks to the media coverage they have been getting, and have been the subject of some of Filipović's academic writings.

In a paper titled "Three Bugs in the City: Urban Ecology and Multispecies Relationality in Postsocialist Belgrade," Filipović introduces several insects (among them that stinkbug, and also the Asian ladybird beetle, *Harmonia axyridis*) that have recently colonized his postsocialist city. In this city, since the demise of the Socialist Federal Republic of Yugoslavia and the lifting of the Iron Curtain, neoliberalism has led to seemingly uncontrolled import and export, and neglect of public infrastructure from the bygone socialist era. That and investor-driven, poorly planned private buildings, he says, stimulate the spread of exotic species.

He has a point. Economically highly lucrative types of commerce, namely the agriculture, pet and aquarium, and gardening trades, depend entirely on the shuttling of animal and plant species across the world, usually to places where they do not occur naturally. The hubs for these trades, the airports and seaports that usually are located in or near cities, are wells of plenty from which an endless stream of non-native plant and animal species flows forth. Some, such as garden plants, are released into the urban environment intentionally; others, like that stinkbug, accidentally hitch rides on traded produce. Unless strict controls are placed on imports (a very un-neoliberal thing to do!), the city will become a hotspot for exotic species, which, since they have left their natural enemies and parasites behind in the place where they came from, can multiply unfettered.

Moreover, the postsocialist forces in Belgrade may lead to a decline of the native biodiversity that could have kept exotic species under control. The public greenspaces that harbor birds, spiders, and other potential predators of introduced insect species are under threat, as they usually sit on prime land for investors. North of the city center, for example, a beautiful quiet area snakes around the Kalemegdan fortress and along the right bank of the Danube. Public parks, sports fields, a railway line, unplanned vegetation ("an oasis for bees," says Jovana Bila Dubaić, a local entomologist), and a Roma settlement lie shoulder to shoulder, and in the summer, many an overheated Belgradian escapes the sweltering city center and takes a stroll to cool down and commune with nature. But such unmanaged, unprofitable public spaces do not sit easily with city authorities, so the city has begun to transform the area in the Belgrade Linear Park with a series of development projects to include high-end apartment buildings, a marina, and tidy, manicured parks, all of which, ecologists fear, will permanently damage the local urban biodiversity.

Yet Filipović points out that, rather than admit that the absence of checks on imports and the degraded urban environment give these exotic species free rein, the authorities much prefer to vilify the immigrants themselves. He analyzed news reports about exotic insects in Belgrade and discovered that media outlets, radio and TV programs, and government-issued reports had three features in common.

First, instead of using the neutral terms biologists use to describe the demography of species, newspapers would talk about the increase in numbers in these insects in overtly militaristic terms: Belgrade was said to be "under attack" from these "armies of invaders," which "occupy" the city and "kill the natives." Second, a strong focus was placed on the Asian origin of the stinkbug, the ladybird beetle, and other exotic species, usually printed in capitals, with a generous helping of exclamation marks: "STINK BUGS FROM CHINA ATTACK!" read one popular tabloid headline. Finally, says Filipović, the media were bent on instilling what he calls "biofear" by hammering on the harm these insects can do to humans, whether true or not. "An interviewee told a popular news website that she was woken 'by a sharp pain in the eye and . . . saw a ladybird beetle on the pillow. It was orange, smaller than ours, with more black spots. I had to take antibiotics because my whole eye was swollen.'" And the webzine Mondo ran the headline, "Invasion of Stink Bugs in Serbia, Some Suck Blood!" (They do not.)

I hasten to say that, though this account stems from Serbia, the situation is similar in most other countries. Driven by the same causes as in Belgrade, the influx of exotic species into urban areas is on the rise worldwide, and the general discourse about them features the same characteristics: militaristic metaphors, biofear mongering, and an appeal to xenophobia. And while it is true that some non-native species, as they disperse without bringing with them any of their specialized natural enemies, can disrupt ecosystems, especially those that are ecologically compromised—islands or lakes with low biodiversity and simple, vulnerable food webs, for example—there is nothing intrinsically evil about them. So, to quote two headlines from Dutch newspapers, why would we say that the kingfisher (as if it were a long-lost friend) is "making a comeback," but that the ring-necked parakeet (as if it were an enemy army) "is advancing," when both colorful bird species enjoy an identical population increase? The only difference between them is their native or non-native status—and such a status is arbitrary in itself: 15,000 years ago there were no kingfishers yet in ice-age Holland, and several million years ago there were native parrots here.

Still, as Filipović points out, words and framing are powerful tools. Either consciously or subconsciously, the notion that exotic species are bad, dangerous, and unwanted nestles in the minds of the general public. To take an example from my own street: for decades, an Asian tree of heaven (*Ailanthus altissima*) had been growing in the square behind my house. It was nice and tall and leafy, and its crown, rustling in the midday breeze, provided shade in summer and nesting opportunities to common wood pigeons (*Columba palumbus*) and magpies (*Pica pica*) in spring. Snails and springtails fed on the lichens and algae that grew on its bark, and a wide variety of beetles, wild bees and hoverflies fed on its pollen and nectar. In winter, I would sieve rich communities of rove beetles and pillbugs from the mounds of its fallen leaves that the autumn storms had piled up in the alley between the houses. Yes, it was an invasive tree, but it was *our* invasive tree.

Until one day I came home from a trip and it was gone. Three men from a "tree management company," in coveralls and helmets and with heavy equipment lying around, were busy loading the sawn-off trunk onto a truck, chopping up the larger branches, and sweeping up and bagging the thinner ones. One of our neighbors looked on with folded arms and a satisfied expression on her face. "I called the municipality yesterday and they came immediately," she beamed. "I read in the newspaper how dangerous these exotic trees are. They will replace it with a native tree." For a moment I felt like my eleven-year-old self again, standing by my rabbit herb garden, sprayed with herbicides.

And, indeed, since then a spindly linden sapling has been languishing in the old tree of heaven's place. It does not look happy. The snails, birds, pillbugs, and insects ignore it, and it will take decades before it has a chance of shouldering all the ecosystem services that the tree of heaven provided. Whose interest has been served by this intrusion into the urban ecosystem? Probably mainly those of the companies that do the exotic species management and of the authorities, who are seen to "do something." The alternative, an ecologically much less damaging laissez-faire policy, could easily be interpreted by the general public as the authorities not doing their job.

In the meantime, as if in revenge, fresh seedlings of "our" tree of heaven are rapidly sprouting all over the neighborhood, for *Ailanthus altissima* really is a very well-adapted urban tree. The city with its urban heat island and poor, disturbed soils is its perfect niche.

URBAN IMMIGRANTS

Even if a campaign of hatred is being waged against exotic species, fueled by authorities' desire to deflect attention away from their own failed environmental policies and feeding into the general public's sensitivity to ecological xenophobia, surely science has a more neutral stance? How does ecology view the roles that exotic species can play?

First, it is crucial to make a distinction for isolated habitats. Real islands in the ocean, and any other place that has been, in an ecological sense, an island for a long time (think of ancient lakes or mountaintops), often have a unique "endemic" biodiversity. Many are populated by small communities of species that have evolved there from the few ancestors that managed to reach it. This makes island ecosystems vulnerable. They are full of unique species that live only there and have not been tested against the regular influx of competitors that is such a central feature of life on the continent.

The ecosystems of such insular places are easily thrown in disarray by exotic species. Many are the examples of islands where the irreplaceable, home-grown food web has been overrun and permanently changed by a single non-native species. The stories are as numerous as they are tragic: the Nile perch (*Lates niloticus*, introduced for fisheries), which ate its way through an entire community of hundreds of endemic cichlid fish species in Lake Victoria; the brown tree snake (*Boiga irregularis*, in the middle of the twentieth century a regular stowaway on Australian cargo planes), which drove to extinction thirteen native bird species and two bat species on Guam; and the rosy wolf snail (*Euglandina rosea*, a predatory snail that eats other snails), which was released in French Polynesia to get rid of the exotic giant African snail (*Lissachatina fulica*) but instead ate forty-eight local *Partula* snail species into oblivion, to name but a few.

While the ecosystems of islands are home-grown, self-assembled over long periods of time in quiet isolation, and therefore easily unhinged by uninvited exotic species, cities are the complete opposite. The restlessness of human endeavor means that the urban environment is in a constant state of upheaval. The frenzy of building and infrastructure projects, the ceaseless rivers of commuters in and out of the urban conglomerations, and their hubs of cutthroat trade mean that organisms are being shuttled in and out of cities like there is no tomorrow. Patches of vegetation come and go, and entire mini-ecosystems—the multitudes of organisms in soil, planting material, and freshwater—are being moved around constantly. The urban ecosystem is a very dynamic place, a melting pot in which exotic species are playing an ever more prominent role. With so many immigrants that can do the ecological jobs as well as or even better than the natives, the urban ecosystem is more than just an analogy of the diverse, multiethnic community of humans that lives there.

The health of the urban ecosystem, like that of its human equivalent, is not served by unchecked xenophobic sentiments. The ecological frameworks of many large cities worldwide are run largely by immigrant species. In the San Francisco Bay and Delta, the communities of snails, sea stars, crabs, worms, and sea urchins that live in the muddy bottom, the plankton in its brackish waters, and the fishes in the freshwater of the rivers that flow into it consist 40 to 100 percent of exotic species. In cities in central Chile, of the more than three hundred species of trees and shrubs that grow wild across urban centers, only forty are native; the rest are introduced species from all over the world. And in the city of Dunedin, New Zealand, there is a one in two chance that a bird you spot belongs to an alien species.

Of course, these novel urban ecosystems, cobbled together haphazardly from whatever species became available, cannot be compared with the delicately balanced ecosystems of rainforests and other natural environments, where species have had millions of years to coevolve with one another. But, says Ingo Kowarik of the Technical University in Berlin, Germany, these urban ecosystems, albeit willy-nilly and despite all their bumpy ecological interactions, *do* work. "Novel mixtures of alien and native species that emerge after profound habitat transformation on urban land are expected to be adapted to novel environmental conditions and in some cases possibly better adapted than previous assemblages of native species," he writes in a landmark paper published in 2011 in the journal *Environmental Pollution*. "[They] may be valued as an adaptation to severe habitat transformation and may secure the provision of ecosystem services in urban settings in an era of global change."

Moreover, their topsy-turviness may wane over time as the species in such mixed native-exotic ecosystems gradually adapt to each other, either by learning to recognize one another or even by evolving rapidly, as we shall see in a later chapter. And community science is proving very important in documenting such changes.

COOKING CRAYFISH

In Mexico City, users of the platforms iNaturalist and eBird took hundreds of pictures of hummingbirds sipping nectar from flowers. Ecologist Oscar

Marín-Gomez from the National Autonomous University of Mexico used all those photos to work out the ecological network among seventeen hummingbird species and eighty-six flowering plant species. He and his team found that the hummingbirds were making full use of exotic plant species. Depending on the type of urban habitat, the hummingbirds were particularly partial to sticking their tongues into the flowers of African klip dagga (*Leonotis nepetifolia*), Australian crimson bottlebrush, *Melaleuca citrina*, South American tree tobacco (*Nicotiana glauca*), and bougainvillea, also from South America.

In the UK, back in 2010, the researchers Michael Pocock and Darren Evans got into cahoots with a nationwide team of primary schoolchildren to study part of the ecosystem that is supported by horse chestnut trees. The horse chestnut (*Aesculus hippocastanum*) is a large, chunky, handsome tree. This exotic species originally hails from the mountains of the Balkans and Turkey, but it has been planted in cities worldwide. In the UK, its springtime ears of pink or white flowers and its large, freshly green, hand-shaped summer leaves have been a familiar feature of large streets and avenues since the seventeenth century. In 2002, however, that all changed. In that year, a small moth from one of the chestnut tree's homelands, probably North Macedonia, arrived in the UK. The caterpillars of this horse chestnut leaf-miner (*Cameraria ohridella*) burrow through the leaves, causing them to become blemished with lots and lots of brown spots. These telltale signals of its presence were first picked up in London but spread across the country fast: by 2010 brown-spotted leaves were seen almost everywhere in England and Wales.

Guided by their teachers and trained volunteers, Pocock and Evans's team of schoolchildren collected brown-spotted horse chestnut leaves, put them in sealed plastic bags, and then waited for two weeks to see what would happen. What happened was that, after a while, the caterpillars would pupate inside their leaf blotches and emerge as tiny moths with pretty auburn-, black-, and white-striped wings. But also some small, brown, black, or metallic wasps would emerge: native parasitic wasps whose mothers had pierced the leaf with their egg-laying device and injected eggs into the moth caterpillars, after which the wasp larvae had eaten the caterpillars from the inside out and emerged victorious instead of their moth victims.

What Pocock and Evans discovered when they did the number crunching of all the wasps tallied by the schoolchildren was that the rate at which parasitic wasps attacked the caterpillars was much higher in areas where the moth had been present for longer. In other words, the horse chestnut leaf-miner is in the process of becoming integrated into the ecosystem of insects that live in and around the trees. The longer the moth has been around, the more time there has been for parasitic wasps to adapt to this new type of victim.

The cause behind the ever-increasing embrace of this exotic moth by its native enemies is probably that it takes time for a tree-based ecosystem of parasitic wasps to build up. These wasps are generalized parasites: tracing them by their acute sense of smell, they attack a wide range of small caterpillar species. Before 2002, the leaves of the horse chestnut did not give off any caterpillar scent, so the wasps would not be interested in them. Once this changed, the tiny wasps would emerge from horse chestnut leaves and probably overwinter behind loose bark or in the dead wood of the tree itself, leading to an even greater attack rate the next year, and so on. So the incorporation of the exotic moth species into the native ecosystem is probably thanks to its enemies' demography.

Other exotic species are incorporated into the urban ecosystem thanks to learning, observing, and imitating. While it is known that parasitic wasps can learn, birds are much better at picking up new tricks. In the canals of my hometown of Leiden, several bird species, such as the coots of the previous chapter but also gray herons (*Ardea cinerea*) and great crested grebes (*Podiceps cristatus*), have learned to catch and eat the invasive American red swamp crayfish (*Procambarus clarkii*) that live in the water by their hundreds of thousands. According to my neighbor Daaf Sloos, lesser black-backed gulls (*Larus fuscus*) have even discovered that, in summer, they can roast the crustaceans by dropping them on the hot metal roof of his shed and leaving them there for a while before eating them. And Bram Koese, who masterminded the roadkill project with the white crosses that I talked about a few chapters ago, is using photos of waterfowl with crayfish in their beaks that he finds on community science platforms such as Observation International to investigate the rate by which these invasive species are being adopted into the birds' menu.

So exotic species can be adopted into a native ecosystem thanks to demography and learning, but there is also evolution. The arrival of a new exotic species in the urban ecosystem causes natural selection that makes some resident species rapidly change genetically—*evolve*—in response. Several years ago, for example, my students and I discovered how, in the Netherlands, the leaf beetle *Gonioctena quinquepunctata*, which previously fed only on native rowan trees (*Sorbus aucuparia*), quite suddenly began to feed on the exotic American bird cherry (*Prunus serotina*). Carefully dissecting the beetle's genome revealed the telltale signs that it is in the process of rapidly adapting to this new food source. And probably the same is happening with other native insects discovering exotic plants in other nooks and crannies of the urban ecosystem.

What I want to say is that urban ecosystems are ecologically very different places from pristine natural environments. The rules and practice of nature conservation have been developed in such wild places but do not always apply to cities equally well. In short, the urban ecosystem is not a place where we should be combating exotic species. Rather, we should embrace them as essential components of the churning, fomenting mix of ingredients that together make up the novel ecosystem that is evolving among us in our cities.

The next two chapters take a closer look at the "evolving" aspects of the urban natural world. Do animals and plants really embark on new evolutionary paths as they are naturally selected by the unusual mix of demands that city life places on them? You bet they do!

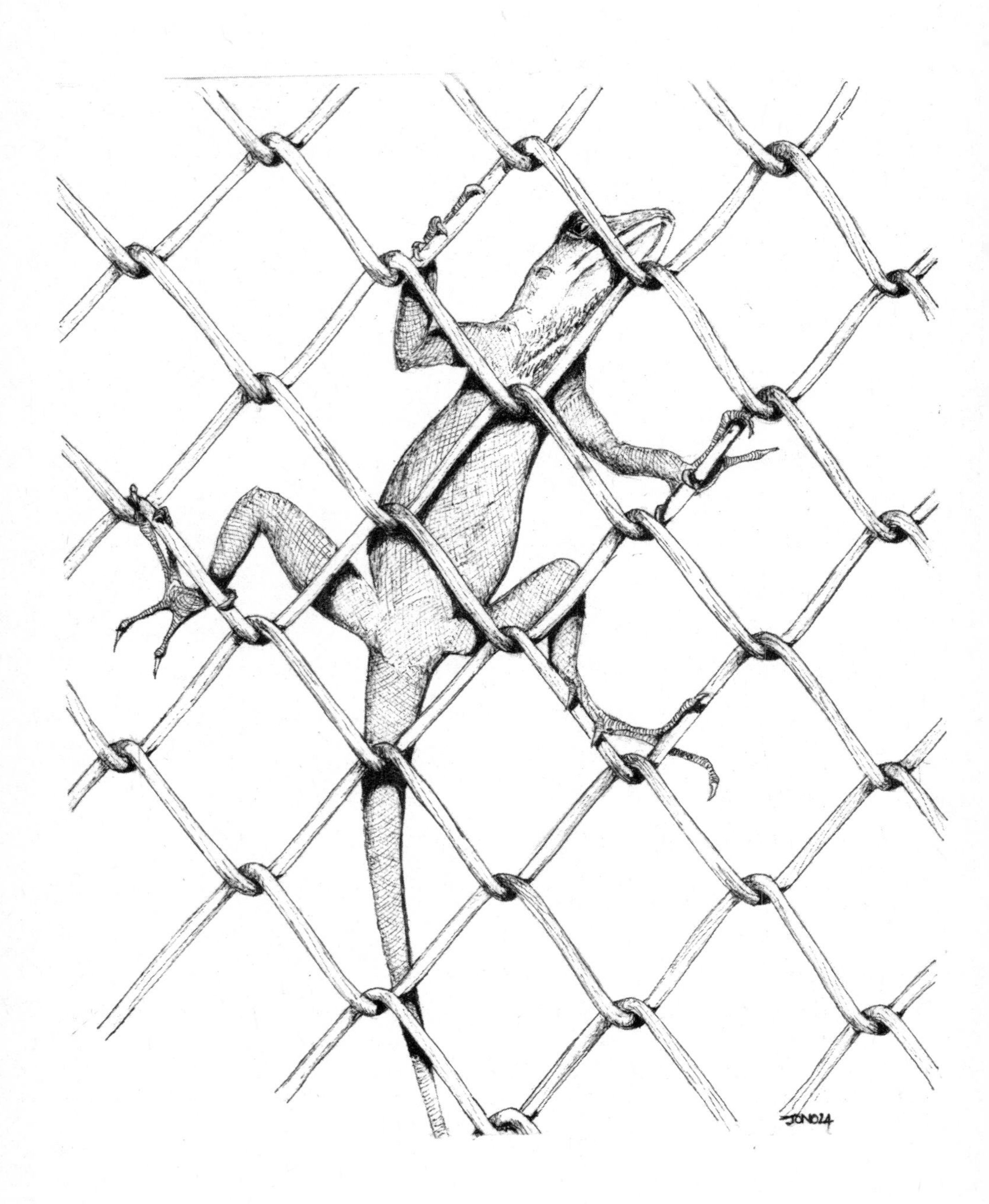

15

ON THE ORIGIN OF URBAN SPECIES

"There is nothing like reaching into someone's yard with a ten-foot-long pole and then having to explain to them that you are not in fact trying to steal from their house, but rather, you want to catch a lizard that is hanging out in their planter," says Cleo Falvey, a PhD student at Rutgers University in Camden, New Jersey. What she is describing is her fieldwork as an undergraduate student at the University of Massachusetts. The lizard in question was an *Anolis*, one of nearly five hundred she caught; the place was a residential neighborhood in the city of Arecibo, Puerto Rico.

Colloquially called "anoles," these sleek, attractive, multicolored lizards are everywhere in the Americas. Anoles are conspicuous because they are common and active and come in a multitude of species, each with its own food and habitat preferences. They are, in fact, a pinnacle of vertebrate evolution, beating such species-rich groups as bats, mice, and warblers. With over 425 species, *Anolis* has more diversity than you can shake a stick at. In fact, shaking sticks (with a floss lasso affixed to the end) at them is exactly what generations of biologists have been doing for more than a century, catching them and studying them, as the pretty lizards and the question of their evolutionary prowess have become a favored problem to sink their scientific teeth into.

In recent years, that line of research has been taking a detour into cities, and that is where Cleo Falvey comes in. She was part of a team led by Kristin Winchell at New York University that has been discovering subtle changes in *Anolis* lizards that live in cities. Falvey, for example, has been catching lizards belonging to five different *Anolis* species on four Caribbean islands, both in cities and in natural forests, and measuring their toenails. Yes, their toenails. She catches a lizard and then holds up the largest front claw (always the third digit) and takes a picture of it exactly from the side. Then she repeats the process with the longest hind claw (always the fourth digit).

Another undergraduate student, Bailey Howell, now doing a PhD at Virginia Tech in Blackburg, while working in Puerto Rico on the species *Anolis cristatellus* did something similar a little bit higher up the lizard limb: she looked at toepads. Just south of the claws that Falvey studied, anoles have a broadened section (the pad) under their toes that is beset with miniature, delicate, parallel skin flaps, or lamellae, making the underside look like a microscopic washboard. A few years earlier, Winchell had caught (and released) over a thousand *A. cristatellus* in cities and forests in Puerto Rico. Each caught lizard she gently placed on a flatbed scanner and, using the highest resolution possible, produced a detailed picture of the underside of their feet. (Imagine, for a moment, the bewilderment in those lizards' eyes as they were made to sit still on the glass plate while the scanner lamp slowly crawled underneath them, illuminating their undersides from one end to the other.) Howell's job was to take measurements of the shape and number of those toepad lamellae.

What Falvey and Howell found (their results were published in the *Biological Journal of the Linnean Society* and *Integrative Organismal Biology* in 2020 and 2022, respectively) was that city anoles have different feet from forest anoles of the same species. Urban lizards' toepads are longer and larger (5.6 square millimeters in the forest, nearly 6.9 square millimeters in the city) and have on average one more lamella. The lamellae are also more widely spaced. The toenails in the city anoles are different, too: shorter, straighter, and wider than in the forest anoles.

The cause? Evolution by parkour. Forest anoles run around on branches of trees and vines, chasing insects. That is what they have been doing for millions of years, and that is the niche their feet are adapted to. But when they entered the city it was like walking into a freshly mopped bathroom on dancing shoes. No longer were they moving around on rough natural surfaces but running and leaping among smooth, slippery human-made ones: windowsills, fence posts, polished marble or enameled tiles, rain pipes, and gutters. In the century or so that they have lived an urban life, natural selection caused by lizards fatally slipping from tall buildings has made evolution think on its feet, as it were, coming up with rapid improvements. Broader toepads with more lamellae give a better grip. And those shorter toenails, Falvey thinks, may actually help give the toepad a better grip on a smooth, hard surface where a long nail (in contrast to a nail pressing into the bark of a tree) would only get in the way.

Since Winchell's earlier research showed that the shape of the *Anolis* foot is largely genetically determined, what we have here is real breakneck-speed evolution, the urban kind that my previous book, *Darwin Comes to Town*, is all about. As we saw in previous chapters, the urban environment stands in stark ecological contrast to natural environments, with its own peculiarities and special challenges but also opportunities. One important aspect is simply the physics of the city: urbanized areas are hotter, noisier, and covered in impervious surfaces such as brick, concrete, glass, and metal. It is the texture of those surfaces that urban anoles are adapting to. In fact, by digitally analyzing photographs of the places where the lizards hung out, Falvey determined their "surface roughness" and showed that indeed, urban perches are twice as smooth as the ones in the forest.

For urban plants, it is not the impervious surface itself that poses a challenge but rather the fewness of the remaining patches of soil for them to take root in. Are flowers and seeds of plants perhaps evolving to adapt to such a plant-unfriendly environment? In 2012, urban ecologist Arathi Seshadri from Colorado State University tested this idea with the common dandelion, *Taraxacum officinale*. What Arathi discovered was that the seeds of dandelions growing in the city (which are suspended under that familiar umbrella-like fluffy bit) float down to ground level a bit faster than dandelion seeds from meadows outside the city. This helps dandelions survive in the stony urban environment where patches of bare soil are few and far between. Think about it: fine as it may be in a meadow to cast your seeds far and wide, in the city, most of those will land on pavement or tarmac and not germinate, except the lucky few that land in the same small patch of soil where the mother plant grows—which is something you can achieve if your seeds drop so fast that they cannot get blown very far.

But Arathi could not work out exactly how the urban dandelion manages this feat; her measurements showed that the "handle" of the umbrella was a little shorter in the city, but how this made the seeds fall faster remained a mystery for the next ten years, which is when my colleague Barbara Gravendeel, an urban evolutionary biologist, tried her hand at continuing where Arathi had left off. In three cities in the Netherlands, her student Chrysoula Karetta determined the rate of descent of hundreds of dandelion seeds. She also took detailed measurements of the size and shape of the seeds and the umbrella under which they are suspended.

These studies confirmed Arathi's results that urban dandelion seeds drop to the floor faster than rural ones. But Karetta and Gravendeel also figured out how this worked on a physical level. What seems to happen is that the dandelion simply moves "building materials" from the umbrella into the seed. As a result, the seed gets heavier, the ribs of the umbrella get sparser, and the handle of the umbrella gets shorter. The end result is a heavier seed hanging from a more tattered umbrella that drops to the ground faster, and therefore has a greater chance of landing in the same patch of soil where the mother plant grows. And these advantages are inherited by the next generation: when the researchers planted urban and rural seeds in the same soil in the lab, the lab-grown plants showed the same seed differences as their parents in the wild had.

Before Karetta started her project, Gravendeel and postdoc Niels Kerstes first attempted to tackle the dandelion project with citizen science. They did a series of experiments with visitors at our natural history museum that involved dropping and measuring dandelion seeds. Blowing the umbrella'd seeds off a dandelion's flower head is a childhood game everybody loves and continues to do well into old age, so it seemed an obvious urban evolution community scientist experiment. Still, as Gravendeel and Kerstes recall, they learned some valuable lessons. For example, while it was easy to persuade small children to drop a seed, it proved much harder to get them to then pick it up again and drop the same seed multiple times, which was needed to get a more accurate measurement. Their attitude was, drop something once, why drop it again? Also, gently picking up a seed without damaging the umbrella was something smaller children did not have the motoric finesse for, resulting in many, many squashed, squeezed, and mangled seeds, unfit for redropping. Grownups posed their own set of problems: they were always in a rush to check out the rest of the museum or the restaurant. Also, with so many people moving around, the air was rarely still, which did not help in getting accurate measurements.

The patchy vegetation that makes dandelions in cities evolve is a small-scale pattern, but cities display such green patchiness at any spatial scale. In a way, it is a fractal pattern that shows up no matter how you zoom in or out on the map of a city: from the centimeter scale (patches of moss or lichens on roof tiles and windowsills), through the intermediate scales at which dandelions experience it or the road islands that we saw earlier in this book, to the scale of hundreds of meters at which the city parks sprinkled among vast built-up areas normally manifest. It is this largest-scale archipelago of green fragments at which larger, mobile animals react to the islandlike qualities of the city's vegetation. One such animal is the Australian water dragon (*Intellagama lesueurii*), a large, heavily built, spiny lizard, a reptile that has Oz written all over. With its greenish gray dark-banded skin and facial and throat decoration of white, yellow, red, and black, it looks like a soldier in battle fatigues who has just returned from a children's party where there was a face painter. Males can be over one meter long, and because they live semiaquatic lives, they superficially resemble the famous marine iguanas of the Galápagos Islands. And that is appropriate because the parks that dot the city of Brisbane are the archipelago in which these lizards are playing out their own evolutionary game.

A team of Australian researchers led by Celine Frère and Bethan Littleford-Colquhoun of the University of the Sunshine Coast in Queensland took it upon themselves to figure out whether the water dragons in the urban parks of Brisbane's Central Business District are evolving in the same way as animals on real islands. For we tend to associate evolution with archipelagoes. Think of Hawaii or other insular worlds, places where many animal species have evolved simply because they have got stuck on separate islands, which forced each of them to embark on its own evolutionary path. Could something similar be already happening to native water dragons marooned in these islandlike city parks, which are much younger than oceanic islands, having been swallowed up by the expanding city since the mid-nineteenth century?

To figure that out, Littleford-Colquhoun and her colleagues took DNA samples from nearly three hundred dragons from four Brisbane parks, and also from some three hundred from six continuous natural habitats outside the city. They measured the jaw, limbs, and tail of each of the animals. What they found (and published in *Molecular Ecology* in 2017) was that these Brisbane dragons indeed are the urban answer to the Galápagos. The lizards in each of the parks formed a genetically distinct population, four times more different from each other than the populations in the outback were, even though those populations were mostly further removed from each other than the distance separating the city parks (which were just a couple of kilometers apart). The same was true for their looks. The dragons in Roma Street Parklands, for example, were small, with broad jaws and short limbs, whereas those in Brisbane City Botanic Gardens, just over a kilometer away, were large, with narrow jaws, long front legs, and short hindlegs. Again, such differences were not seen between the dragon populations in the wild.

What this means is that these lizard populations, stuck in their own little park for dozens of lizard generations, have been evolving at lightning speed. Although the researchers cannot yet put their finger on the precise scenarios, it probably was the specific conditions in each park (the types of available food, how much human-made surface was available, the amount of water, the numbers of trees) that the lizards have adapted to. Incidentally, this discovery also feeds into the single large or several small (SLoSS) debate that we talked about earlier. With each dragon population adapted to its own little patch of urban nature, combining those patches into one large park, or connecting them with green corridors, would create a situation in which it would be more difficult for the smaller local populations to maintain that genetic uniqueness: they would become swamped and "genetically polluted" by immigration from the larger parks.

EVOLUTION IN A NUTSHELL

Ozzie water dragons, Caribbean anoles, and Dutch dandelions may be teaching us how the urban ecosystem is evolving in response to humans clothing cities in impervious surfaces, but there are other physical features of city life to reckon with. One of those is the so-called urban heat island. Millions of people and their machines crammed into a small area filled with buildings that block the wind and soak up the sun's heat is what causes cities to be hotter than the countryside. In the center of a large city it can easily be 6 or 7°C warmer than at the same time outside that city. This means that city centers are literally hotspots embedded in a cooler landscape. Thermally speaking, you could think of, say, Los Angeles as a piece of central Mexico inserted into California, and London as a chunk of southern France transposed into England.

Urban heat is something that has a considerable effect on urban ecology because temperature is one of the leading factors determining the trials of life of wild organisms. This is especially true for so-called sessile organisms, creatures that are stuck in one spot and cannot actively seek out cooler spots. Plants, of course, are the ultimate sessile organisms, but some animals are pretty sessile, too—land snails, for example. For most of my career, I have been working with land snails precisely for that reason. In the wild, they live up to their proverbially tardy reputation: more than once I have done experiments in which I wrote numbers on their shells and visited the same spot a year later (many snails are surprisingly long-lived) only to discover that most were sitting within a few meters from the spot where I had marked them the previous year. For an evolutionary biologist, they are wonderful animals to work with because their slowness means they will evolve to adapt closely to the local, often harsh, conditions—unable as they are to move around and seek out more benign conditions as many other, more mobile, animals do.

In the city, evolving because of the local heat island is something you would expect snails to do. So, in 2014, postdocs Niels Kerstes and Thijmen Breeschoten, my colleague Vincent Kalkman, and I set up a research project to see if we could detect this in the brown-lipped snail, *Cepaea nemoralis*. The brown-lipped snail is the unsung hero of evolutionary genetics, having been studied since the late nineteenth century by generations of biologists. This has to do with the pretty color patterns they have on their shell. The shell's ground color can be a fresh lemony yellow, a bright pink or orange, or a deep chocolate brown, and it may be adorned by up to five parallel black bands that spiral along with the shell's whorls. And, like humans' hair or eye color, the hues and shades of *Cepaea*'s shell are purely genetic. There is one gene that determines the ground color and over a dozen genes that manipulate the black bands: turning on or off specific bands, changing their width and their regularity.

These shell color patterns are one way in which the snails defend themselves against higher temperatures. Experiments have shown that the snails with the lightest shell color (pale yellow and no bands on the top part of the shell) are better protected against overheating than ones with a darker shell (with a darker ground color and more black bands). In summer, snails often have to survive for days or weeks glued to a wall or a tree trunk, exposed to the full sun. For the same reason that it stays cooler inside a white car than inside a black one, the body temperature of a light-shelled snail remains several degrees cooler than that of a dark-shelled snail, and this may mean the difference between life and death for a summertime snail.

So we hypothesized that the thermal natural selection acting against snails with dark shells in city centers would lead to the evolution of lighter urban shells compared with shells of snails in rural areas around the city. To test whether our hunch was correct, we set up a classic citizen science experiment. And by classic I mean it was a project in which trained scientists called the shots and citizen scientists were mainly involved in collecting the data—a type of citizen science that we have since abandoned in favor of projects in which the citizen (or community) scientists take a more leading role. In this case, we built a simple point-and-shoot smartphone app (named SnailSnap) that allowed people to snap pictures of urban and nonurban snails. The pictures were automatically uploaded to the citizen science platform Observation International (also used for some of the roadkill projects that we saw earlier in this book), where a team of ten volunteers sorted them by color variant.

Altogether, individual citizen scientists, nature clubs, and schools in the Netherlands sent in nearly ten thousand snail pictures. The number crunching that Kerstes did on the data after the photographs had been classified by our team of volunteers showed that, sure enough, the urban snails were, on average, lighter (more likely to be yellow) and were less likely to have black bands on the top of the shell. Our hunch had been correct!

Though at first sight more mobile than snails, urban ants find themselves in a similar predicament. A team of researchers led by Sarah Diamond at Case Western University in Cleveland has been looking at a tiny species of urban ant called *Temnothorax curvispinosus*. These ants, of which the workers have a body length of only two millimeters, are so small that the entire colony (a queen, up to a hundred workers, eggs, larvae, and pupae) fits in a single acorn shell—hence the informal name of acorn ant. A queen colonizes the empty husk of an old, fallen acorn and rules over her nut-sized kingdom for up to five years. Although older workers leave the nest on short foraging trips, most of the colony is stuck in their miniature home—pretty much like a snail in a shell.

And, like snails, acorn ants too feel the urban heat. Diamond's team picked up altogether twenty-four acorn ant nests in urban and rural areas in and around Cleveland (which forms an urban heat island around 4°C hotter than the countryside), let them acclimatize to room temperature in the lab, and then tested the temperature tolerance of the workers. They did this by putting a worker into a small vial, placing the vial into a heater, and then slowly ramping up or down the temperature until the worker showed loss of muscular coordination or, as Diamond calls it, a kind of "drunken staggering." In other words, as soon as the worker got so hot or so cold that it could not keep itself upright anymore, the temperature was recorded and the experiment was stopped. For the experimental subjects it was too late, though: "The thermal tolerance assays were destructive, preventing the return of these worker subsets to the colony," the authors write wryly in their 2021 paper in *Evolution*.

Nonetheless, the ants did not die in vain, for the experiments showed that the urban worker ants' tolerance for both cold and heat had shifted (evolved) by about one to two degrees Celsius compared with rural ones. That is, urban worker ants stayed alive until about 47°C, while rural workers had already gone belly-up by 46°C. At the lower end of the temperature range, the difference was similar: at a frosty 4°C the rural workers could still function, while the urban ants had already chilled into immobility at temperatures well above 5°C.

Now comes the nice part. A favorite way for researchers to test whether animals have evolved to adapt to their local environment is to do a transplantation experiment: moving animals from their home range into a new environment to see how they fare. Normally, this is very laborious: you need to transplant lots of animals and carefully keep track of their individual fates in their new homes. But with these acorn ants, it was a piece of cake: all the researchers needed to do was move the ant colony–containing acorns from their rural homes to an urban one, and vice versa. As a crucial control, the colony-containing acorns were also transported from one urban home to another urban home and from one rural place to another rural place to test the effect of the move itself, without any change in ambient temperature.

They did this for groups of ten acorns and then followed the survival of these colonies over nearly two and a half years, censusing the acorns eleven times, at one- to four-month intervals, by picking them up again from their new homes, transferring them briefly to the lab to count the numbers of inhabitants, and then bringing them back again. By the final census, in August 2019, it had become clear that rural colonies in an urban environment had mostly died out, and, importantly, they mostly died during summer. Urban colonies transplanted to a rural environment also did not survive well, but they predominantly perished in winter, a pattern consistent with the temperature tolerances of the urban and rural workers that had evolved to match the urban heat island and the cooler rural conditions, respectively.

Only a little bit off where heat sits on the electromagnetic radiation spectrum lies visible light. And, like heat, light is another physical aspect of the urban ecosystem that has been cranked up considerably compared with wild, nonanthropogenic ecosystems. Meet ALAN.

ALAN

What do anonymous graffiti artists (with tags like "GRAM" and "Twa") and the tiny spider *Brigittea civica* have in common? Only that they are both among the chief culprits accused of defiling the beautiful porticoes of the Italian city of Turin. Or at least that is how the Centro Conservazione e Restauro, the city's art conservation and restoration center, views them. In its program to combat "urban degradation," the center has been tackling what it considers to be the main causes for the aesthetic decay of Turin's famed centuries-old porticoes. Measuring in total eighteen kilometers, second in length only to those of Bologna, these grandiose colonnades connect streets and squares, allowing pedestrians, vendors, and the patrons of café and restaurant terraces to remain dry in winter and cool in summer. But their columns and vaulted ceilings are also favored blank canvases for the city's graffiti artists (or, as they are called on the program's website, "graphic vandals") and also for an important member of Italy's urban ecosystem, the appropriately named civic meshweaver, *Brigittea civica* (for which the Centro Conservazione e Restauro has not yet coined a derogatory slur).

These spiders, with a body only three millimeters long, build small circular webs about five centimeters in diameter. Their webs are not sticky: mesh web weavers are so-called cribellate spiders: they fray their silk into a multitude of fine threads that catch prey by entangling their tiny feet rather than by gluing them down. But cribellate silk is also very good at catching dust particles from the air, which is why the initially invisible webs in Turin's archways quickly darken into dirty gray blemishes on the white plaster. Add to that the fact that mesh web weavers are social spiders (they like to build their webs close to each other, even overlapping, and share prey without any ill feelings) and you can understand their bad reputation among Turin's art restoration authorities. The ceilings of some parts of the ancient arcades are covered in hundreds of grayish brown cobwebs—eye candy for the urban ecologist, and a potential biological air pollution monitoring system, but an eyesore to the architecturalist.

While themselves dealing with the graphic vandals in a separate subproject, the Centro Conservazione e Restauro authorities subcontracted a team of arachnologists at the University of Turin to study the offensive spiders. Led by cave spider specialist Stefano Mammola, the team's project to map the spider webs in the cavernous archways of Turin's porticoes was not too far removed from its usual speleological field expeditions. They set to work in the colonnades of Via Po and Piazza Vittorio Veneto, which cover a distance of about two kilometers. Based on their initial surveys, they suspected a positive influence on cobweb density of artificial light at night (also known by the acronym ALAN) from the old incandescent and more modern halogen light bulbs placed in the porticoes. To be able to measure that light pollution, they worked only at night. In seventy-two arcades, at three moments throughout the year, they photographed a random, 0.7-square-meter section of ceiling and counted the numbers of *Brigittea civica* spider webs on their images. For each section, they also measured the amount of light shed onto that section by the nearby lamps.

They found that the cobweb density was indeed strongly dependent on the artificial lighting, especially from the old incandescent lights. In summer, when the spiders and their prey were most active, the web density would be around six times higher close to an incandescent light compared to places that were several meters removed from such a light source. In some well-lit places, there was almost no webless space left on the walls. For the halogen lamps, the influence was much less. So one of their recommendations to the authorities was to renovate the lighting in the arcades and use a type of light that the spiders seemed less attracted by.

Superficially, because that is also what their prey do, it may seem obvious that spiders flock to light sources. However, accepted arachnological wisdom says this should not be so. In contrast to moths and other flying insects, spiders are actually light shunners. Time and again they have been shown, at least in wild, natural habitats, to be lucifugous—that is, to be *repelled* by ALAN, not attracted to it.

And yet Turin's civic meshweaver is not the first urban spider that has been seen to build its webs close to light sources. Already in the late 1990s in Vienna the same was recorded in bridge spiders (*Larinioides sclopetarius*). On a pedestrian bridge across the Danube Canal, the zoologist Astrid Heiling discovered that the spiderwebs were strongly concentrated around the light bulbs that illuminated the bridge, and that these webs-near-lamps also caught up to four times more prey than webs in dark places. Anybody who has walked around a city and paid attention to where spiderwebs are will have noticed the same. So could it be that urban spiders have evolved to do away with their ancestral fear of light and have instead adopted a new light-seeking behavior, as a result of the higher prey capture that their webs would be able to get when close to a source of ALAN?

That is exactly the question that Tomer Czaczkes of the University of Regensburg, Germany, and his colleagues wanted to answer with a laboratory experiment. They focused on yet a different species of spider, the triangulate cobweb spider, *Steatoda triangulosa*. In southern Germany, northern Italy, and southern France, they found these spiders in four urban locations and two rural ones. But rather than catching adult spiders and checking in the lab whether the spiders would prefer to build their web near light sources or not, they decided to work with newborn spiders. The adults, after all, could somehow have *learned* that light sources offered rich pickings, while newborns would be naïve, so any difference among them would need to be innate, and therefore probably the result of urban evolution.

In the triangulate cobweb spider, fortunately, it is quite easy to obtain baby spiders because the spiders package on average thirty eggs into each of the fluffy silken egg sacks that they guard in their web. So the researchers needed only to snatch egg sacks from spider webs, bring them to the lab, and wait for the spiderlings to hatch. In all, from these six sites, they managed to get nearly eight hundred healthy baby spiders this way. Meanwhile, they prepared their experimental setup. In the lab, they installed eighty identical plastic boxes. Each box had two connected compartments and a window on one side; a small LED light was installed on the outside close to that window. In this way the boxes offered a bright and a dark compartment. Within two days after being born, individual baby spiders were released into an experimental box, and after forty-eight hours the researchers checked in which compartment the spiderling had decided to build its very first web: in the dark or the light section.

As it turned out, the rural spiders still adhered to the traditional habit of spiders to shun the light: they mostly built their webs in the dark compartments. But the urban spiders chose the bright compartment twice as frequently as their unenlightened rural brethren. In fact, while there was no evidence that they *preferred* the light compartment, they definitely had lost their fear of it, and they had also lost their preference for darkness.

You may not have noticed it, but with *Steatoda* we have turned a corner in our tour of urban creatures evolving in response to the physical fabric of the city. The parkour anoles and umbrella-seed dandelions were adapting to hard surfaces, the water dragons to the fragmentation of greenspaces caused by these hard surfaces, and the snails and acorn ants, holed up in their shells, to the urban heat island. The factors that make these organisms evolve are all static in the sense that they are not affected by whether or not organisms adapt to them. But with spiders losing their fear for ALAN, it is a different story. Yes, ALAN is a physical feature, but remember, spiders are not adapting to ALAN itself but rather to the fact that their prey (moths and other flying urban insects) are attracted to

ALAN. And these insects *can* evolve themselves. In other words, ALAN creates an arena wherein the *interaction* between two or more urban species is changed: as spiders evolve to be more attracted to light, the insects they prey on might, in response, evolve a fear of light.

Swiss researcher Florian Altermatt found precisely that. In both lit-up downtown Zurich and in the dark countryside outside the city, he caught hundreds of caterpillars of the small ermine moth, *Yponomeuta cagnagella*. He then brought these to his lab, kept them under identical conditions, and waited for them to pupate and emerge as moths. Then, after giving the moths a paint mark so that he could tell urban from rural ones, he released them in mixed bunches in a large mesh cage that had a single light trap at the end, and counted how many of the moths ended up in the trap. As it turned out, the urban moths' tendency to fly to the light was reduced by about 30 percent compared to the rural moths' tendency. In other words, urban moths had evolved to be somewhat immune to ALAN.

Now, there are many disadvantages for moths to fly to light besides ending up in a light-loving spider's web. They might get burned or singed by the heat of the light, or they might spend the whole night sitting close to the light, transfixed by its brilliance, but also wasting time that they should have spent searching for food or for mates. Still, Altermatt's experiment (which he and his colleague Dieter Ebert published in *Biology Letters* in 2016) shows how urban evolution in a predator-prey system could become a dance in which every evolutionary step taken by the predator is answered by a counterstep from the prey.

The next chapter looks at a few more examples of urban evolution where the thing driving the evolution is not physical or chemical but a living, breathing, and potentially coevolving urban coinhabitant.

16

WE ARE A NODE

Ask an AI program to come up with an image of an assassin in a sushi bar and you will get scenes of mean-looking ninjas looming over displays of nigiri and sashimi, but chances are very small that it will create something like the view I used to witness on a daily basis when I lived in Kota Kinabalu, the capital of the Malaysian state of Sabah. My writing desk was placed on the first floor of my house in Likas, one of the older residential neighborhoods in this city of half a million. It faced a wall and a window. Watching columns of pharaoh ants (*Monomorium pharaonis*) marching through gaps in the frame of that window was one of my favorite ways to procrastinate as I sat writing at that desk.

The tiny ants, even smaller than the acorn ants we saw earlier, nested in secluded spots in my house; I would often come across their nests in boxes of CDs or even in piles of laundry that had been left in one place for more than a few days. But for foraging they preferred to go outside, raiding the pile of kitchen refuse (potato peels, fish heads, cheese crusts) in the yard. A four-lane carriageway of worker ants would permanently supply the nests with food. The route had been established so long ago and was so heavily used that the tip-tap of their millions of soiled feet had left a faint but discernible gray streak that snaked from the window across the whitewashed walls and into the darker corners of the house where their nests were.

But I was not the only one observing their comings and goings with fascination. In front of me, perched on the edge of the window frame, just millimeters away from the ant highway's hard shoulder, there would be an assassin bug, *Vesbius purpureus*: a sleek creature on thin, black, stilt-like legs wearing a black-and-scarlet cloak in which it hid a long, sharp, daggerlike proboscis. Whenever an ant wandered too close, the bug would flash it out in an instance, pierce the ant's abdomen, lift it off its feet, and suck it dry. Then it would use its front legs to remove the empty carcass off its snout and get ready to impale its next victim. It would sit there nearly motionless for hours, feeding with minimum effort as the sushi train of ants kept bringing in new fresh food.

But *Vesbius purpureus* was not the endpoint in the food chain that channeled nutrients from my compost heap via the pharaoh ants. The assassin itself was assassinated on a regular basis by the flat-tailed house geckoes (*Hemidactylus platyurus*) that clung in all manner of position to the walls and ceilings of my house, constantly on the watch for a tasty insect or spider to present itself to their insatiable appetite and add it to the multitude of dry, elongated, black lizard droppings with the white dot of uric acid on one end that are such a familiar if mixed blessing of living in the tropics.

The geckoes themselves were a delight to watch as well. They would often get into ceiling-bound skirmishes in which one or more became dislodged and would belly-flop onto the polished wooden floor, sit still in shock for a few seconds, and then scurry off in reptilian embarrassment. Their social interactions as well as their chasing the many insects that had been attracted to the porch light would often lead them outside the house as well, where they now and then fell prey to the brilliant blue-and-white collared kingfishers (*Todiramphus chloris*), always perched on the electricity wires and always on the lookout for easy pickings, which they would descend on with that maniacal cackle that echoed through the neighborhood.

So here we have a food chain in which nutrients from my kitchen compost heap were carried back into my house by the pharaoh ants and then absorbed by assassin bugs; the bugs themselves contributed nutrients to my house geckoes, which in turn donated their bodies to kingfishers. And although I never observed it, I'm sure the kingfishers would occasionally also be eaten by a predator, perhaps while they were distracted in the act of slapping a freshly caught and still resisting gecko against the rocks of the retaining wall—maybe by a passing Brahminy kite (*Haliastur indus*), large white-and-tan raptors that often cruised in the air above our street, or, at night, by one of the black spitting cobras (*Naja sumatrana*) that lived in the pipes of my retaining wall.

And this five-species food chain is itself just a small snippet from a much larger urban food web. For the ants are food not just to assassin bugs but to many other predators and parasites. And the assassin bugs, although specialized on feeding on ants, probably also preyed on any of the other species of ant that cohabited with me in my tropical town-house. In many ways, such urban food webs are very similar to food webs in other habitats—complex networks of who eats whom. But there is one crucial difference: *Homo sapiens*.

The energy input that makes ecosystems in natural environments tick is mostly locally produced by the sunlight-powered plants that turn water and CO_2 from the air into the food that feeds the rest of the food web. But think of the food chain that I just described: it was powered not by the green plants in my garden but by kitchen refuse. And where did that kitchen refuse ultimately come from? From places much further afield. The potato peels came from farms on the cool slopes of Mount Kinabalu, a hundred kilometers north of the city; the fish heads from the South China Sea, probably hundreds of kilometers west; and the crusts of ched-dar from New Zealand, made from the milk of cows grazing thousands of kilometers away. In other words, the urban ecosystem is heavily subsi-dized by food that *Homo sapiens* imports into the city—and much of that human food spills over into the urban food web of wild creatures.

What this also means is that we humans are an important node in the network of ecological relationships in the city. In fact, we are a keystone species, without which large parts of the urban ecosystem would collapse. Directly or indirectly, intentionally or involuntarily, our food feeds many other species in the natural fabric of the city. This may be in such an obvious way as somebody feeding bread to the ducks in a park pond or in much more discreet, invisible ways such as sewage leaks that feed the city's soil microorganisms. Either way, the omnipresent *Homo sapiens* is a central species in the maintenance of urban biodiversity. Removed from nature as we may think we are in our glitzy urban environment, we are actually much more entwined in it than in most nonurban places.

This also means that the interactions between humans and other urban species are fodder for evolution. Since we are such a vital source of nutrients, evolution will look kindly on any characteristic that will help a species pilfer a larger portion of that human-based food supply.

In North America, the masked bandits with their dexterous paws called *Procyon lotor* (raccoons) are in an arms race with trash can designers, being able to open more and more complex locks. In Australia, sulfur-crested cockatoos (*Cacatua galerita*) have figured out how to open household waste bins, and by imitation and innovation this skill has swept through the cockatoo populations of Sydney, Wollongong, and beyond. But although the ability to solve puzzles and to learn has a genetic basis in birds and mammals, these behaviors have spread so quickly that evolution by natural selection probably has not played a role.

For another prominent animal from the Sydney urban core, evolution *does* seem to be changing its personality and behavior. It is that chunky lizard of Oz, the Australian water dragon (*Intellagama lesueurii*) that we met in a previous chapter. Not surprisingly for urban lizards, their diet is full of human foodstuff. James Baxter-Gilbert, who studied these animals for his doctorate at Macquarie University, writes in his PhD thesis, "I have observed them consuming a variety of cooked human foods (e.g., hamburger, hot chips, and popcorn) [and] pet foods. . . . In public areas such as picnic sites, they are known to actively approach humans and scavenge food dropped or thrown to the ground." So one of the things Baxter-Gilbert wanted to know is whether living among humans has changed the lizards' personality. Are they bolder, more curious, or more outgoing than their rural counterparts? In wild environments, it often pays to be circumspect and wary. But in cities, with the abundance of food that can be obtained from humans, fortune may favor the bold.

To test this idea, Baxter-Gilbert chose the same method as Tomer Czaczkes used with his cobweb spiders: circumventing the possibility that an animal's behavior is due to nurture rather than nature, he cut out the nurture part by working with freshly laid eggs. He caught egg-laden female water dragons in various places in and around Sydney, brought them to his lab, and let them lay eggs there. Then he incubated all the eggs under the same lab conditions: identical temperature, humidity, what have you. Once the baby lizards hatched, each was chipped and then released in a lab enclosure. The enclosures were all the same: large plastic tubs arranged side by side, each filled with sand, a few tiles, a log, and a small pond.

Every two months, Baxter-Gilbert subjected the nearly one hundred adolescent recruits to a battery of personality tests. For example, he would release them into a new tank and record how much time they spent exploring these new surroundings; this showed how keen they were to explore unfamiliar grounds. Boldness was assessed by measuring how long it took for a lizard to recover after a stressful experience. The experience in question was chasing it around with a blue-gloved hand until the animal felt compelled to hide in an unpleasantly cool place of the enclosure. Then he measured how much time elapsed before the lizard dared show itself again and seek out a warmer spot. Finally, Baxter-Gilbert studied the dragons' neophilia, or curiosity about novel objects. He placed a typical urban item (an empty coffee cup, an unopened soda can, an unopened bag of potato chips) into the enclosure and measured how close the animal dared to approach it within half an hour.

The measurements showed that lizards from urban environments were no more curious or explorative than the ones from wild bushland, but Baxter-Gilbert did pick up a difference in boldness. After their unnerving blue-gloved-hand encounter, urban lizards came out of hiding nearly ten minutes sooner than the ones from wild environments, who refused to show themselves for up to forty-five minutes. Since the lizards had all grown up under the same conditions, they differed only in whether their mum had been caught in an urban or a nonurban environment. Hence the difference in boldness seemed to be innate and perhaps the result of rapid evolution, spurred on by encounters with humans putting a premium on temerity.

The water dragon results fit into a pattern of personality evolution in other urban animals that regularly interact with humans. Urban birds of a variety of species, for example, are more tolerant of people than members of the same bird species living in rural places. They allow people to come closer to them and they are more curious about human-made objects.

MUTANT MOZZIES

But some species in the urban web of ecological interactions depend on humans in a more sinister way—mosquitoes, for example. Think of the opportunities that a megalopolis chock-full of the exposed skins of potential blood donors offers to these parasites. In cities in the warmer parts of the globe, one of the dominant mosquitoes that feed on human blood is

Aedes aegypti. This insect is originally from tropical Africa but is now found as far afield as India, Southeast Asia, northern Australia, Latin America, and the Gulf Coast states, and has established outposts in southern Spain, Philadelphia, northern California, the Middle East, and Tokyo. That is worrying because *A. aegypti* is not just a buzzing nuisance, it also transmits the human diseases dengue, zika, chikungunya, and yellow fever.

The bloody connections in this corner of the urban food web that is run by *Homo sapiens* blood flowing into the *A. aegypti* gut sets the stage for mutual evolution—because, let us not forget, there is a third player on this stage: the virus that exploits the body of its host, causing disease and death in both human and insect. And the sad reality of evolution is that natural selection thrives when death and disease are rife.

Let us look at the mosquito first. Many large cities in Africa are home to dense populations of *A. aegypti* mosquitoes that feed exclusively on humans. This might seem obvious, given that human blood is the most abundantly available type of blood around. But most other species of *Aedes*, and also the populations of *A. aegypti* that are found in rural areas, prefer to bite birds and nonhuman mammals. In fact, when Noah Rose of Princeton University and his colleagues looked at mosquitoes from twenty-three different locations across tropical Africa, they discovered that there are populations of the mosquito that go for human blood and detest any other type, but also the reverse, and everything in between. (They tested this with a so-called Y-tube experiment: place a mosquito female in the bottom leg of the glass tube Y, insert an experimenter's arm in one of the two upper legs and a guinea pig in the other, and see where the mosquito flies to. Rinse and repeat.)

When Rose then looked at the environments from which his team had obtained the twenty-three mosquito populations, he found two factors that strongly determined their propensity to bite humans: one, the local density of people, and two, the severity of the dry season. He also discovered that all the human-biting mosquitoes shared the same seven regions in their chromosomes. In other words, probably as cities emerged in Africa over the past millennia, the mosquitoes had evolved a preference for humans because of the double Darwinian whammy of lots of available blood to feed on and lots of available urban water reservoirs for their larvae to live in during the dry season.

The viruses, meanwhile (yellow fever, dengue, and zika are all closely related members of the genus *Flavivirus*), are also evolving to suit the multiple demands of having temporary homes inside both humans and insects. When inside the mosquito, they multiply fast enough for sufficient virus particles to be passed from mosquito to human during the brief moment that a mosquito sucks in human blood; but at the same time, suffering from the virus should not make the insect too sick to fly. Once in the human, the virus should do everything to get into the blood stream in high numbers while keeping its human alive and well long enough to be visited by hungry mosquitoes to continue to pass on the virus. In other words, urban *Flavivirus* evolution constitutes a balancing act among all these competing requirements.

Humans, finally, the third partner in this eco-evo-triangle, are hampered in their evolution by the fact that they have such a long time span between being born and having children. Since natural selection works with a time lag (only in the next generation do you see the effects of natural selection that affect the current generation), humans are in evolution's slow lane, being overtaken left and right by viruses and mosquitoes, which have much shorter generation times and therefore a much higher evolutionary clock speed. Still, human geneticists such as Marisa Oliveira of the Cambridge Institute for Medical Research are finding telltale signs of urban human populations evolving resistance against dengue. For instance, the gene *OSBPL10* (short for oxysterol binding protein-like No. 10) has evolved slight differences in people from densely populated areas of Africa. These differences make the gene less active, which seems to make it harder for the dengue virus to enter human cells. In other words, an evolutionary countermeasure against the disease's greater impact brought about by *A. aegypti*'s greater predilection for human blood has emerged.

THE EVOLVING CITY-PERSON

Humans evolving resistance against the dengue virus may be just one of the many ways in which the urban ecosystem makes humans evolve just as much as other creatures. In 2013, Joséphine Daub, then at the University of Bern, Switzerland, published a paper in *Molecular Biology and Evolution* in which she and her colleagues compared the genomes of over a thousand people from all over the world to look for mutations that were suspiciously common in certain parts of the world but rare elsewhere, or the other way around—a sure sign that there had been a recent sweep of evolution affecting the gene in which the mutation took place. They found hundreds of such genes. More important, they also found that most of those were in some way or another involved in the immune system, and concluded that they must have been affected by the sudden settling of people in dense conglomerations, and the new opportunities for the spread of diseases that this offered. (The recent COVID-19 pandemic has brought home their point in an apocalyptic manner.) A few years later in the same journal a Russian-Swiss team compared human DNA from archaeological digs all over Europe with that of modern humans living in the same place and also concluded that Europeans' immune systems, among many other things, had evolved since urbanization had kicked in.

We often hear it claimed that modernity has made humans stop evolving. With culture and technological and medical progress protecting us from every ill that befell our ancestors, life expectancies having multiplied, and with child mortality having dwindled to a fraction of what it was, natural selection in our species must surely have come to a grinding halt—or so the reasoning goes. And yet there are many reasons to believe that the reverse is true, and that today we are in fact witnessing a crucial jolt to the evolution of our species.

First, we are creating better and better conditions for evolution to take place: genetic mutations are a chance process, and with eight billion people on Earth, the appearance of improbably rare, beneficial mutations has become a near certainty, while the ethnic melting pots that cities are prevent such mutations from remaining stuck in isolated populations. Second, the fact that couples in modern urban societies have fewer children than during pre-industrial times makes natural selection potentially stronger rather than weaker, since the impact on the next generation of a two-child family is double that of a one-child family (a much greater difference than between having, say, nine versus eleven surviving children). And third, the urban ecosystem, even though we created it ourselves, is as drastic a change from our natural environment as it is for any urban animal or plant. In fact, using a genetic data set called HapMap, a team of researchers calculated that the rate of evolutionary change of our genomes today is between ten and a hundred times greater than it was 40,000 years ago.

Of course, the urban evolution in humans that has been taking place since the onset of urbanization is not visible in the same way that the transition from *Homo erectus* to *Homo sapiens* is visible. We are not talking about gait or brow ridges but rather about subtle physiological changes that only show up when we begin comparing genomes (but that, when accumulated over a long time, *will* create a new human). And today, with the cost of reading an entire genome still falling (it has been hovering around U.S. $500 since 2019, down from a whopping $10 million as recently as 2007), and with computing and statistical power on the rise, detecting ongoing human evolution is getting easier and easier, as recent publications show. Those publications offer fascinating insights into the kind of evolution we are talking about.

In Mexico, for example, the frequency of the e4 variant of the Alzheimer gene APOE is higher in rural areas than in the city. This is because e4 is advantageous in places with poor nutrition as it prevents diarrhea in children. But in areas with energy-rich foods full of cholesterol, it is harmful and causes heart disease and Alzheimer's in the elderly.

And the UK's Biobank, which holds genetic data on half a million Britons, yielded information that showed that a variant of the gene CHRNA3 that predisposes people to nicotine addiction declined over the twentieth century as a result of chain smokers dying prematurely from lung cancer. Studies using the Biobank data also revealed that genes that determine that women have their first baby earlier in life and a later onset of menopause (in other words, genes that lengthen the reproductive part of their lives) have increased over recent decades. Under more stable, urban conditions, in which food security allows most children to survive, women who raise more offspring obviously have an edge in contributing their genes to the next generation.

There we have it: the urban ecosystem is a complex network of all urban life forms, with humans as a keystone species at the hub of it. It is also a network of new opportunities and new risks to which each and every urban species, humans included, is adapting by rapid evolution. So, rather than viewing the city as a biologically depauperate, uninteresting place where nature is degraded by the hideous and detrimental activities of us humans, we might see it as offering an an exciting new opportunity to watch evolution in action. The creation of a globally distributed novel ecosystem, especially one ruled by a single species (us), is a pretty rare event in the history of life on Earth, and we are here to see it happen. And by "we," I do not mean just evolutionary biologists like myself. Many of the urban evolution phenomena that I have described in this chapter could be revealed by community scientists as the experiments needed are ridiculously simple. The evolution of ALAN avoidance by urban moths is an example. Not to downplay the efforts of Florian Altermatt, but anybody, including a team of dedicated community scientists, could have reared caterpillars and let them fly toward the light. And yet, in the nearly ten years since Altermatt published his paper, nobody has tried to replicate his study. "It is so obvious that I am surprised too," Altermatt wrote to me in an email.

And so I hope to end this chapter with a call to arms. Go out into your local neighborhood and pick an urban evolution phenomenon to study. We professional scientists are too few and have too little time to do all the work. If I take a walk in the old inner-city streets around my house in Leiden, I see so many urban animals and plants that are just begging to be studied. In between the cobblestones grow smooth rupturewort plants (*Herniaria glabra*); have they perhaps evolved a shorter stature or a different flowering period after having been trodden on for centuries by urban pedestrians? The bush crickets (*Tettigonia viridissima*) whose shrill papery song is heard from the old plane trees, do they sing at the same pitch among the noisy traffic as their rural relatives? And what about the sawflies (*Tomostethus nigritus*) that are suddenly appearing everywhere on the cultivated "Raywood" ash trees that the municipality has planted: have they adapted to using this garden-center tree as their host plant? Each of these questions can be tackled by simple, good, old-fashioned experiments that any urban naturalist with a bit of ingenuity and scientific curiosity can carry out in their kitchen or their garden shed. It will help you understand and appreciate that urban ecosystem of which we all are an integral part. It is an ecosystem that is worth conserving and safeguarding against all the threats it faces in the fast-paced, densely populated urban world. Urban nature conservation, therefore, is what the final part of this book is about.

KNOWLEDGE IS POWER: THE URBAN NATURALIST-CONSERVATIONIST

To be effective, our volunteers, our citizen conservationists, must be committed. To be committed, they must believe. But . . . the citizen-conservationist must first understand it in order to believe it.

—Michael Rosenzweig, *Win-Win Ecology* (2003)

No tract of land is too small for the wilderness idea.

—Aldo Leopold, *Wilderness Values* (1942)

17

SPEAK SOFTLY AND CARRY A BIG STICK INSECT

The police inspector's eyes, just inches away from mine, are one of the most haunting memories of my childhood. Or actually, the memory is not so much of the eyes themselves but rather of the bags under them, which I inspected in great detail as he was shouting at me. They were fleshy, purplish brown, and contracted and expanded in concert with the narrowing and widening of his eyes as the scolding he gave me moved through its series of theatrical routines. I was particularly fascinated by the two horizontal lines that I discerned among the wrinkles. The skin there was smoother and paler, and seemed less flexible. I wondered what had caused them. They were too neat and symmetric to have been the result of some injury suffered in the course of police work. My conclusion that this macho, bulky policeman had undergone cosmetic surgery helped me remain proud and quiet while I underwent his telling off. The only two words of his tirade that I committed to memory were the mocking "environmental activists!" spoken with a mixture of disgust and derision. I had heard the same words uttered in the same tone at family gatherings. A traditional lineage of farmers from villages to the north of Rotterdam, my mother's side of the family had enjoyed great fertility around the turn of the twentieth century and, at birthdays and weddings, would embed themselves in large numbers in the plush sofas that my mother would arrange in a circle along the walls of the living room and drink jenever, smoke cigars, and contentedly dismiss all the progressive novelties that the mid-1970s came up with. "Environmental activists? Criminals and communists, the lot of them!"

I, the only child at these parties, would sit quietly among my great-uncles and great-aunts, happily helping myself to the cheese cubes with mustard, mini-gherkins, and roasted cashew nuts that my mother had put out and, with occasional nods and smiles, would give the assurance that I too thought the hippies of Greenpeace should be locked away for good. As soon as my parents indicated I had attended the party for the required amount of time, I would withdraw to my room upstairs and read my bird books or arrange my collection of dried skulls and shells. Or do my homework: drawing diagrams of the ecosystem of the Wadden Sea, the endangered estuary in the north of the Netherlands that my hippy, Greenpeace-supporting schoolteacher devoted many of his most popular lessons to.

So when my friend Isaac and I discovered that the ditch near our house, a paradise of clear water where we used to study water beetles and catch dragonflies and sticklebacks, had been turned into a bluish, milky dead zone because of mysterious discharges of the clubhouse of the soccer club, we hatched a secret plan of sabotage. By systematically destroying expensive property of the club, we figured, we would force them into bankruptcy, the club would pack their bags, and the pollution would stop. Our strategy was as simple as it was misguided—after all, we were twelve-year-olds, not seasoned guerrilla activists.

For a while, everything went according to plan. It was winter, and the sun had already set when we came home from school. Under cover of darkness, Isaac and I would sneak into the grounds of the soccer club and see what damage we could do. I would use my Swiss army knife to slash the nets and the dug-out tarps, and Isaac, who had a knack for electronics, would sabotage the switchboxes of the scoreboards and the floodlights. We kept a list of the estimated financial consequences, and our prognosis was that we would need to keep this up for a few months and then the soccer club would be out of business and the ditch would be ours again.

Had we kept track of the local newspapers, we would have realized that our sense of security was as false as it was naïve. Today I can search the digital archives of the town's newspaper and watch our operation's downfall unfold in a series of news items throughout the winter of that year: 21st of December: "soccer club reports damages to the police"; 27th of December: "soccer club says they will start patrolling at night and, if they catch the perpetrators, kill them completely dead"; 29th of December: "more damages"; 4th of January: "two vandals caught and handed over to the police."

And that is how I ended up being shouted at by that police officer. The night before, on our daily spree of destruction, we had walked straight into the arms and baseball bats of the nighttime patrol that the soccer club, unbeknownst to us, had set up. And when we were called into the police station the next day, they did exactly what we had always seen on our favorite TV series: Isaac and I were separately questioned and they used a good cop, bad cop routine to get us to confess. The bad cop was the officer with the scarred eyelids. The good cop was a lady police officer who complimented me on being "tough" for denying everything for the whole duration of the day that they kept me there.

Of course, denying did not help. My parents ended up footing the bill of all the damages, which put our family in dire straits for the rest of the year. I was left with a fear of the police that, for the remainder of my teenage years, had me shuddering at the mere sight of a police officer. My father, whom I adored, made it clear that he was very disappointed in me. And, worst of all, the local newspaper wrote, completely falsely, that our vandalism had been brought on by the fact that our families disliked the noise from the nighttime soccer matches. Not a word about the illegal dumping of wastewater in our favorite ditch (which continued unabated). Not a mention of our noble intentions.

A lot has changed in the nearly fifty years since that police officer spat the words "environmental activists" into my face. In many countries, environmental activism has drifted from the radical fringe into a broadly supported mainstream activity. In my home country, over a quarter of the population is now a member or a registered supporter of Greenpeace, Friends of the Earth, or any of the other environmental groups that sprang up in the 1970s. In local and national politics, environmental parties are sizable or even dominant, and a sports club that would pollute a freshwater habitat in the middle of a suburb with illegally dumped wastewater would immediately be reported by the neighborhood and told to stop by the authorities. Or, more likely, environmentally minded members of the club itself would prevent this from happening in the first place.

And while there is still a long way to go, the numbers of people who care for the environment and put their money where their mouths are, have grown by orders of magnitude.

That is not to say that all is well. Not long ago, a group of activists in my home country climbed into the trees of the Sterrebos, an old forest about to be cut down for the expansion of a car factory. The forest has the legal status of state monument, but it is on the land of the factory, which managed to get that status overturned. For weeks on end, in the middle of winter, the activists camped high up in the trees, adopting the nickname of "bats," but to no avail: on February 8, 2022, the police moved in and violently evicted the human bats, with a battalion of excavators in their wake. In less than a day, the forest was flattened—for nothing, as it turned out, because the factory's expansion was eventually called off because of the economic downturn that arrived just months later. The factory has since closed down.

Still, in an urbanized, densely populated country like the Netherlands, with relatively high incomes and levels of education, such blatant disregard of the value of urban nature and the people who care for it has, fortunately, become much rarer than it was when I was growing up. But in some other countries, urban nature is not blessed with such a large support base. Its survival routinely hangs by a thread, and its defenders routinely risk their lives trying to save it.

GARDENS OF RESISTANCE

Before me lies an undulating pointillist canvas. As my eye hops from nearby to further and further away, the individual buildings merge into a patchwork of subdued earth tones and then, in the far, far distance, where the haze blurs the geometric shapes of the buildings, the cityscape becomes a pale ocher carpet that abuts the forest-clad hills on the horizon. Covering an area of nearly 5,500 square kilometers, this is the largest city of Europe. The abundance of minarets pointing skyward from the blanket of houses already reveals that I am not in Paris, London, or Madrid. Technically, of course, straddled across the Bosporus as it is, Istanbul lies on two continents; but if we consider it a European city, it is by far the largest. With almost 17 million inhabitants, it is roughly twice the size of either London or Paris and holds as many people as the countries of Finland, Denmark, and Norway combined.

As I walk westward across the Şişli district, the Asian and European vibes appear to vie for prominence in concert with the city's topography. Several deep ravines cut north to south through this area, so my walk is going up and down all the time. On the ridges lie broad avenues lined with Parisian-style apartment buildings, but whenever the slopes become steep, the neighborhood reverts to its old Ottoman character with small brick and wooden dwellings covered in vines, steep winding staircases, scattered palms and cotoneasters, and rubbish dumps with crumbling walls covered in pellitory (*Parietaria judaica*), among which creep polluted streams that drain southward into the Golden Horn estuary.

I am making my way to a large green area I spied on my map and that I thought would be a large park but, when I arrive, turns out to be the Feriköy Cemetery. Surrounded by a cast iron fence on which hooded crows are perched, thousands of marble headstones stand under a canopy of cedars, cypresses, and various deciduous trees. The densely arranged graves themselves are raised half a meter above the floor, and each carries a rectangular bed of vegetation; either wild weeds and grasses or cultivated flowers, but all maintained in a strikingly lush state by the young men who walk the otherwise deserted graveyard with watering cans. It is afternoon prayer time, and the muezzins' calls are broadcast from loudspeakers in the surrounding neighborhoods, but for the rest the cemetery is a quiet green island in the middle of an ocean of traffic, noise, buildings, and people.

In fact, says urban ecologist Emrah Çoraman of Istanbul Technical University, the city's extensive graveyards, many of which are ancient and have slowly been absorbed into the rapidly growing city, are among the most important green areas of Istanbul. In a city where space is a much sought-after commodity and the community of concerned naturalists is infinitesimally small, open spaces are sitting ducks both for developers and for the government with its prestigious building projects. The result is that just 4 percent of the city's surface area is devoted to urban greenspaces (compared to, for example, 33 percent in London, or 28 percent in Seoul). Only respect for the dead and fierce protest can preserve a green area here.

If any protest deserves the moniker of "fierce," it is the one that took place at Gezi Park in 2013. Though remembered internationally as a nationwide revolt against Turkish president Recep Tayyip Erdoğan's repressive politics, it started out as a local, small-scale environmental demonstration. Still reeling from the government's 2012 decision to clear huge swaths of forest (an estimated 2.7 million trees) on the northern edge of the city for a new bridge across the Bosporus and a new international airport, people were shocked to hear that the cherished Gezi Park in the heart of the old city was also slated for destruction, to make way for a traffic hub, a new mosque, and a large apartment building and shopping mall.

On May 28, 2013, the police violently cracked down on a small group of activists who had occupied the park to prevent its annihilation. The unprovoked violence against peaceful demonstrators, the defenselessness of an old and well-loved urban park, the bottled-up frustration at the government's continuing abuse of its authority in environmental and other matters, and iconic media images, such as that of a stoic young protester in a red dress being pepper-sprayed by a policeman, all conspired to whip up public sentiment in favor of the protesters. In a few days the protest grew from just a few dozen people to a crowd of tens of thousands and spread to other cities as well. In all, throughout the summer of 2013, an estimated 2.5 million people were involved in what broadened into the largest antigovernment protests in Turkey in decades. Police tried to disperse the crowds at Gezi Park and the neighboring Taksim Square repeatedly, leading to many deaths, injuries, and arrests.

In the end, says a former inhabitant of the area, Rana Söylemez, thanks to the protests, Gezi Park was spared. Only the mosque was built in a corner of Taksim Square, but this did not affect the green area of the park. But perhaps even more important was that the community that had gathered around the protests used the spirit of optimistic togetherness to move on to do other great things in the neighborhood. Not far from Gezi, for example, was a steep vegetated slope with a magnificent view over the Bosporus and old Constantinople across the Golden Horn. In 2013, says Söylemez, it was a derelict area covered in garbage and litter and also a prime location for the planned construction of a series of four-story buildings with cafés and restaurants. "It is literally the only green space left in the district. The last thing we need in this neighborhood is another place to drink coffee," she says sarcastically. So, sparked by an online call for volunteers, she joined a group of former Gezi Park protesters who wished to change the fate of this greenspace by converting it into a community garden, which they named Roma Bostan.

The word "*bostan*" has the simple meaning of "garden," but it is a term laden with historical and social significance. For millennia, the fields on either side of the old Constantinople city walls have served as community gardens where local people practiced an early form of urban farming, feeding themselves and a large chunk of the city's population with its figs, strawberries, and vegetables; the gardens even helped the citizens live through an eight-year siege in the fourteenth century. The lettuce grown in the *bostan* along the section known as Yedikule is even today still fondly known as Yedikule lettuce, with its lovely buttery leaves. This practice and the sprawling *bostan* continued until the middle of the twentieth century, when the rapid development of the city caught up with it. Today, only small segments of the *bostan* remain, and the farmers constantly battle with the municipality and the police who demolish their sheds, bulldoze their plots, and try to evict them in favor of motorways, buildings, car parks, and the kind of sterile designer parks that dot the modern city.

By opening the new Roma Bostan, the group that Söylemez became part of tapped into this age-old tradition. They cleared the rubbish from the area, planted fruit trees, and terraced the steep slope to grow vegetables. The action group's members were not farmers, so they had to learn everything from scratch. "Those first days, it was amazing to see how many people were there," she says. "Everyone was trying to do something. Amazing to see the solidarity!" Söylemez herself also lacked experience. "I had studied material science and engineering, and I worked in a media agency for a while," she says, "but after Gezi I decided this is not what I want to spend my time with, so I took ecological education, and started making [organic] soaps as a small business, using rosemary and other herbs from the *bostan* to make them."

Slowly, the rest of the neighborhood began to embrace the garden. Residents from a nearby apartment building arranged for a water supply. The wife of the imam who lived in a wooden house right next to the *bostan* asked if she could bring some chickens from her village to show her city-kid grandchildren how to care for poultry and get fresh eggs. Still, gardening was just a means to an end. "Our main aim was never to grow food, it was just a way to bring people together and to start conversations, to create a community," says Söylemez. "What we aimed for was to make an example for everyone that you can defend the greenspaces in your neighborhood by nurturing value and ownership. You do not have to wait for someone to come and do something, you can do it yourself."

When I visit Roma Bostan, it is winter, and there is not much activity. The fruit trees are fast asleep, the vegetable beds are waiting for spring, some cotoneasters lining the street are bearing their heavy bunches of deep red berries, and the chicken sheds of the imam's wife are empty. The ubiquitous feral cats are roaming the land aimlessly, stepping daintily among patches of snow and lots of empty snail shells ("Yes, we had quite a snail problem in the first year," says Söylemez). But what is particularly noteworthy is that, nearly a decade on, there is no sign of those planned four-story buildings, and the view toward the Bosporus is still unimpeded. Söylemez: "Instead of standing in front of the machines to stop the demolition, we said, let's take action before those machines arrive, create what we hope for and then see what happens. And in that sense, we were successful because our voluntary association began a court case and in July 2017 we won. This place is going to stay a greenspace, and not only this area but all the greenspaces in the district."

Still, says Söylemez, being an environmental activist in Turkey is much more dangerous than in many other European countries: in the protests at Gezi Park, for example, though admittedly they evolved into much broader issues than just protection of the park, at least eleven people lost their lives. All the more impressive it is that ordinary neighborhood folk still have the courage to rise up against the authorities to defend their local bit of urban green.

DEFENDERS OF THE SULTAN'S WOODS

Across the Bosporus, on the Anatolian (Asian) side of the city, lies the Validebağ grove, a 350,000-square-meter remnant of native woodland with ancient Judas trees (*Cercis siliquastrum*), oaks (*Quercus robur*), and

Atlas pistachio (*Pistacia atlantica*), some over four hundred years old. The Validebağ grove, however, is completely enclosed, surrounded by busy shopping streets, a mosque, a high school, residential areas, and a modern hospital. It is the only piece of wilderness that remains within the dense artificial fabric of Istanbul, and this, says local oral surgeon and bird-watcher Cihan Babuccu, is partly thanks to the fact that it was an estate of the Ottoman sultans. By the time the sultanate became a republic in the early twentieth century, the city had already fully encircled it. In fact, I have arranged to meet Babuccu at the wooden building at the entrance of the grove that is Sultan Aziz's former hunting lodge, Av Köşkü. He (Babuccu, not the sultan) arrives with a pair of binoculars around his neck and his five-year-old son on his shoulders, and apologizes for having to bring his son to school first—he will be back shortly.

While I wait for him to return, I wander about the mid-nineteenth-century elegant wooden single-story building. It has a timber colonnade all around and colorful paintwork. As Babuccu explains to me when he returns, over the centuries, Validebağ has been managed by various members of the complex Ottoman dynasty. At the end of the eighteenth century, Sultan Selim III began a vineyard there for his mother (*Valide* means "mother" in Turkish). Later, in the first half of the nineteenth century, Bezmiâlem Sultan, wife of Sultan Mahmud II, created a botanical garden, which is why the woodland today is quite rich in exotic tree species. When in 1923 the republic was formed, the area was given to the Ministry of Education, which still manages a few historical buildings on the edge of the grove, and facilitates a storytelling program for schoolchildren.

Babuccu takes me on a stroll through the grove, which is a mixture of forest, orchard, and fields. A small stream runs through it, surrounded by dense reeds and thickets of brambles and hawthorn. We pass a half-open parklike area where large cream-colored feral dogs lie on the grass under majestic trees and look at us with a menacing, proprietary air. 'If they attack us, whatever you do, don't run,' warns Babuccu, who points at other walkers on the paths who carry big sticks with them for this very reason. Toward the east the area slopes up, which affords a pleasant view over the entire woodland.

We stop there, and Babuccu produces his binoculars and points out some of the bird species: magpies (*Pica pica*) flapping from tree to tree, a team of Alexandrine parakeets (*Psittacus eupatria*) zipping by in close formation, and a lesser spotted woodpecker (*Dryobates minor*) flying in the distance in its undulating manner. We listen to European robins (*Erithacus rubecula*) defend their winter territories with their fragile calls of watery trickles and gurgles. During early autumn, this is one of the best spots to watch migrating birds, Babuccu says, as the Bosporus funnels the flyways of many bird species exactly over Validebağ, and they often take a break there to rest and forage. "On some days, there are so many woodlarks, siskins, honey-buzzards, hawks, that I don't know where to look first." In a soft voice that betrays enduring quiet indignation, Babuccu begins to tell the story of all the threats that his urban bird paradise has endured over the years.

Although Validebağ has been officially protected as a nature reserve since 1999, he says, greedy eyes have been ogling it as vacant land just waiting to be "developed." And while some plans (a city forest, "the Hyde Park of Istanbul," a "National Garden," a bird observation tower with walking paths) may sound environmentally friendly, the Validebağ Volunteers and Validebağ Defense NGOs think they are just thinly veiled excuses for money to be earned by well-connected construction companies. They remain categorical in their wish to preserve the area exactly as it is. In recent years, threats to the forest have intensified as struggles over the future of Validebağ have become a microcosm of the power struggles in Istanbul, where the overarching city council is run by the opposition party but some individual districts, including Üsküdar, the one Validebağ is in, are in the hands of Erdoğan's ruling party.

On June 18, 2021, the Turkish minister of environment and urbanization announced the latest planned project for the woodland. In collaboration with the Üsküdar municipality, and seemingly ignoring the grove's protected status, his ministry would carry out an ominous-sounding, U.S. $5 million "Landscaping and Rehabilitation Project." Babuccu says: "They wanted to put artificial grass, artificial playgrounds, and 300 light poles." This was enough for the NGOs to start patrolling the area on a daily (and nightly) basis. On September 21, it paid off. At five o'clock in the morning, two hours before dawn, the municipality arrived in the woodland with trucks, excavators, and twenty police officers. But they had not reckoned with the NGOs' patrollers' WhatsApp network, which quickly brought in a crowd of more than a hundred protesters from the neighborhood. Within half an hour, their human blockade effectively stopped the excavation works. Babuccu: "They asked, why are you opposing us? We are just trying to help, we are making a playground. We said: not necessary; our children already play here!"

For now, their opposition has managed to halt further construction work. But it is a tense stand-off. "We can't trust the government, so we are still on duty," says Babuccu. A large information stall and scoreboard at the woodland's entrance show the number of days that the daily patrols have been going on, and we pass a briskly pacing woman in a thick winter coat wearing a broad armband indicating that she is a member of the patrolling group. She is elderly and respectable-looking—not the young, liberal, fierce kind of activist that I have seen in the photos of the Gezi Park protests. "Exactly," agrees Babuccu. "Many of the NGOs' members are retired, and have spent their childhoods playing in this forest." Babuccu himself is one of the younger members. He runs a busy oral surgery elsewhere in the city, but he makes sure to schedule his appointments only in the afternoons and evenings so that he has the mornings free for birding and volunteering work.

Together with a friend, he has amassed sightings of no fewer than 153 species of bird, nearly a third of the Turkish avifauna. They take beautiful telephoto pictures and post them on the @validebaginkuslari (Birds of Validebağ) Instagram account and also on the citizen science platform eBird, where the forest has its own page. Other volunteers keep track of the plants (over three hundred species) and the butterflies (more than thirty species) and post them on iNaturalist. By showing the people in the neighborhood what interesting animals live in the grove, they are able to further boost the sense of pride of this unique bit of urban wilderness.

Still, says Babuccu, serious naturalists in Istanbul are extremely rare; he estimates their number at a few hundred. For a city of this size, that is a vanishingly small count. Asked why this would be, he explains, "People have to fight for their lives. Fifty percent of people work at minimum wage; they barely earn money. And the stuff you need for this hobby is expensive; if you want to do birding, at least you need a pair of binoculars." But he does notice that the popularity of nature-related hobbies is slowly increasing. Their bird Instagram account chalked up nearly three thousand followers in the one and a half years since they opened it. And more and more people are beginning to send him pictures of their sightings.

Up at Istanbul Technical University, Emrah Çoraman is noticing the same thing. He also estimates that Istanbul contains at most five hundred naturalists, but social media are already helping to increase that number. Making clever use of the university's desire to be known for its dedication to sustainability and biodiversity (which will tick the required boxes to raise its position in the world university rankings), he has been promoting nature conservation on campus. There is a small bit of woodland and a pond on the university's terrain, which Çoraman is highlighting as the campus nature reserve. Also, he is planning to set up a field center there, and to begin a birding group among the students. On Twitter, where he has nearly 50,000 followers, he created the hashtags #hangitür ("which species") and #yabanİstanbul ("wild Istanbul"). These have become hugely popular, and people from all walks of life are using them to highlight pictures of animals and plants they encounter in the city: mostly birds, but also many insects, plants, reptiles, even the fossils people spot in the marble tiles of glitzy modern buildings.

In the end, the power of social media may be what saves urban nature in a city like Istanbul. It is amazing that a simple hashtag on Twitter or Instagram and pictures of cool creatures can stimulate thousands of people all over the city (many of whom may never have had prior interest or the opportunity to pay attention to nature) to begin noticing the animals and plants around them. Occasionally, a post about a particularly cool creature may go viral and reach an even broader circle of people. This happened, for example, with a very rare blanket octopus swimming in the harbor. "It was filmed by someone who shared it in his account," says Çoraman. "When somebody added #hangitür to that post we were notified and it got thousands of likes." And even if most followers of a hashtag lose interest after a while, a proportion may become hooked and turn into true urban naturalists themselves. For the largest city of the Mediterranean Basin, one of the world's thirty-six biodiversity hotspots, this is not too much to hope for.

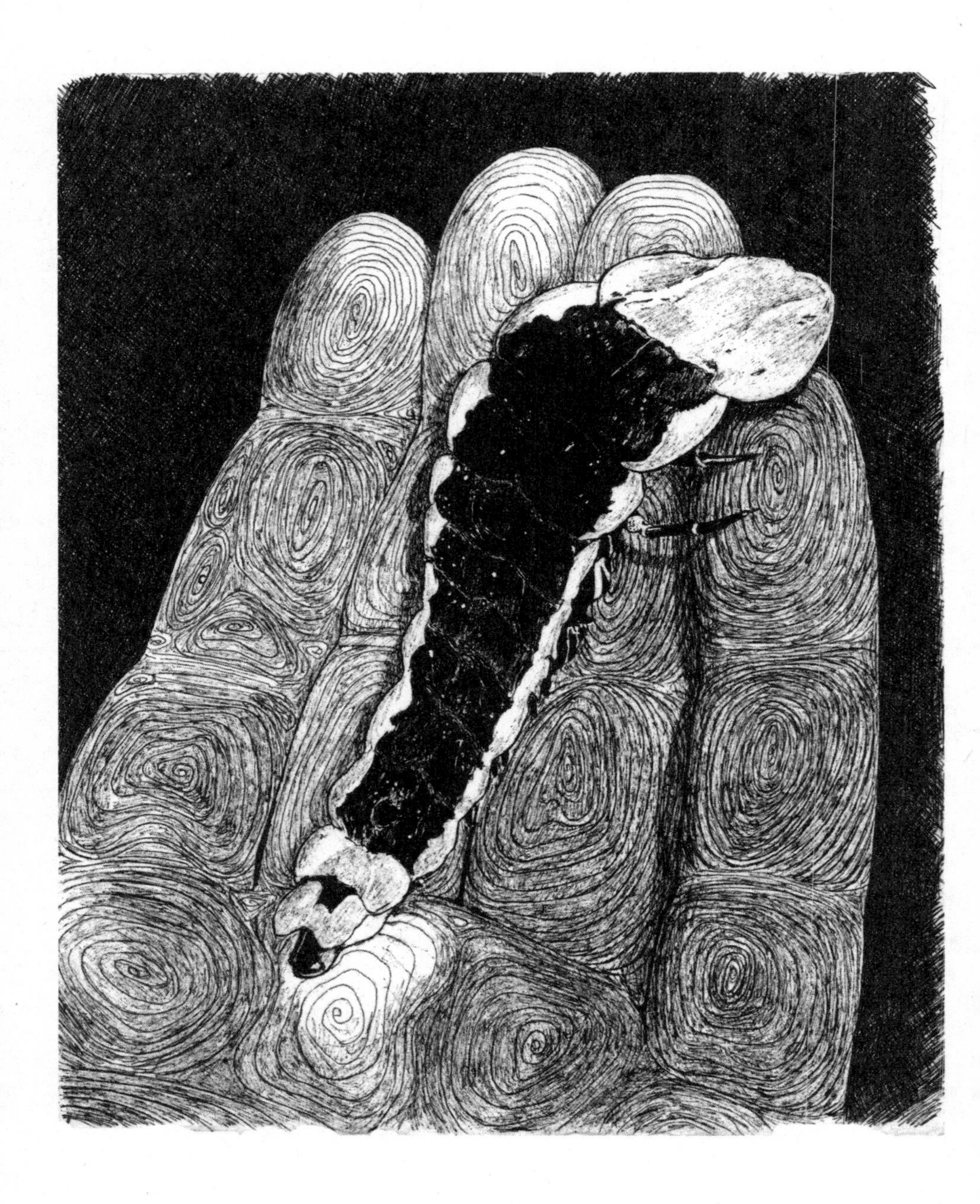

18

TAK KENAL MAKA TAK CINTA

Of all the cities that I have known for a long time, the one that has changed the most is Kuala Lumpur. Back in the late 1980s, when I was a biology student on an overseas traineeship, I spent five months in Universiti Pertanian Malaysia (now Universiti Putra Malaysia), the agricultural university thirty kilometers south of KL (as Malaysia's capital is affectionately known). There, if I was not out in "Ladang 2" tending my corn plants and the parasitic wasps I was using to control corn-eating caterpillars, I was holed up in a hot, moldy concrete student dorm with huge *Periplaneta americana* cockroaches and *Aedes aegypti* tiger mosquitoes for housemates. So each weekend, I would escape campus by catching the mini bus to KL to see a concert in the Old Town Hall, walk around Chinatown, or go to a meeting of the Malaysian Nature Society.

The minibus, having already picked up students at the nearby National University, would already be full, so I would stand in the aisle. Being Dutch and at least a foot taller than most other passengers, and much to their amusement, I could simply prop myself between the floor and the ceiling of the bus and thus keep my hands free to read a book while the bus driver took us through winding roads at breakneck speed. Along the way, we would pass through rubber and oil palm plantations, villages with fluttering road-crossing chickens and lots of motorcycles, along deserted and overgrown open-cast tin mines, and through the endless hardware stores and automobile workshops of Sungai Besi. Only after an hour or so would the bus slow down, merge with traffic from other directions, and, growling reluctantly, enter KL, to come to a hissing stop in the cavernous bus terminal of Pudu Raya.

Today the same trip would go via the modern system of fast, elevated metro lines that now connect all the districts of the city. Along the way, there would be no more villages, chickens, plantations, or deserted mines. In the intervening decades, KL has grown fivefold. The city and its satellite towns have merged into one big metropolitan area of seven million people. The university, complete with its fields where my corn experiment was first eaten by harvester ants and then trampled by a stray herd of water buffaloes, is now embedded into the southern district of the city, wedged in between the new township Putra Permai, multiple north-south and east-west highways, and the glitzy Mines Wellness City, a gigantic urban health resort built on top of one of those old tin mines.

But my old university is not the only thing that got hemmed in by the expanded KL metropolitan area, says nature guide Stephen Wong of RESCU, a KL-based environmental consultancy company. A large forested area of nearly fifty square kilometers used to lie in between KL proper in the east and the residential suburb of Petaling Jaya in the west. But today, both urban areas touch one another almost along their entire length, and what remains of the former forested expanse is now known as the Damansara Arc, a fragile string of green pearls draped across the KL city map.

One of those pearls, Wong says, is Bukit Kiara, or Kiara Hill, a former rubber plantation that was deserted in the early 1980s and since then has been regenerating. Slowly, jump-started by a few small pockets of ancestral forest in the area, the erstwhile monoculture of South American *Hevea brasiliensis* trees has filled up with native flora and fauna. Today it is a wild tropical jungle much loved by the inhabitants of the neighboring residential areas, who go there for hiking, jogging, mountain biking, birdwatching, nature study, and also "hashing" (a quirky game, a relict of the British colonial era, involving a paper trail laid by "hares" and a pack of very determined "hounds" crashing their way through the undergrowth). There is a local Friends of Bukit Kiara (FoBK) association that organizes birding outings, lectures, and night walks.

But FoBK exists for more than nature education and fun events. On its Facebook page, the organization identifies itself as "an environmental rights NGO championing conservation of Bukit Kiara, KL's only urban green lung of substantial size. We seek to build a widely-felt sense of identity & ownership to form a community bulwark against unwanted development." For all is not well with Bukit Kiara. Since the days it gave up being a rubber plantation, it has been gnawed at from all sides, with parcels snapped up by golf courses, country clubs, and property developers.

FoBK, in a series of opinion articles in national newspapers, has been actively highlighting the importance of this urban jungle for the city. In 2007 the government set aside around two square kilometers as a community forest, but since then another 0.25 square kilometers has been lost somewhat mysteriously to various parties. Clearly, the forest needed to become even more famous within KL to prevent it from being devoured entirely by the evil forces of greed. With so many amateur botanists and zoologists active in the forest, there is already an impressive list of flora and fauna recorded there: iNaturalist lists over 1,500 sightings of more than five hundred species of plants and animals. Wong himself spotted some iconic larger animals that have managed to hang on in this relatively small forest fragment, such as the sleek tree-dwelling Asian and small-toothed palm civets (*Paradoxurus hermaphroditus* and *Arctogalidia trivirgata*), the dangerously cute Sunda slow loris (*Nycticebus coucang*), and the ancient-looking Sunda pangolin, or scaly anteater (*Manis javanica*).

And yet the flagship species that the community eventually converged on to serve as an emblem for their beloved forest was not any of the charismatic mammals, colorful birds, or rare snakes that elicit oohs and aahs on FoBK slide show evenings. Going against the conservation marketing dogma that flagship species need to be cute and furry, they chose an insect. And it was not even a spectacular birdwing butterfly or a dazzling metallic jewel beetle but a dark, ground-bound, and rather elusive species: a firefly.

Fireflies are not flies but beetles. In addition to their unique luminescent qualities, other aspects of their walk of life are unusual too. While the males are shaped like a regular beetle (they are the ones that fly), the females are neotenic ("youth retaining"), which means that when as larvae they grow up, they do not pupate and metamorphose into an adult, winged insect. Instead, the larvae keep on molting and growing until they are much larger than the male. Then, at their final molt, they grow ovaries and other female plumbing, a different color pattern, and fancier antennae, but that is about all. To the uninitiated, they still look like very large, flat, maggot-like larvae. Like the males, both females and larvae can emit a greenish yellow light (hence the alternative name of glowworms).

And it was such a larviform firefly female that FoBK member Lim Koon Hup came across one evening in 2017. Granted, it was a very large one, a whopping ten centimeters or so, a member of the genus *Lamprigera*, which comprises the world's largest fireflies. So no wonder it took his eye and spoke to the imagination of the other members of the naturalist community once he shared his sighting with them. Still, it was a leap of faith for them to decide to turn this insect into the forest's emblem and start proclaiming their forest as the home of the world's biggest firefly.

And it worked! A few months later FoBK announced a nighttime firefly excursion, and the event was picked up by a popular KL webzine. To FoBK's bewilderment, instead of the usual fifteen or so diehard naturalists, a crowd of more than five hundred people turned up, all hoping to see "the world's largest firefly." This was an opportunity not to be missed, so, wielding hastily sourced megaphones and rounding up last-minute night-walk guides from among their membership, FoBK introduced a large cross section of KL's populace to the blessings of Bukit Kiara forest and its wildlife, using the *Lamprigera* fireflies (which were obligingly easy to spot that night) as the main attraction.

Spurred on by the success of this first firefly event, FoBK, together with the Urban Biodiversity Initiative of KL-based naturalist Thary Goh, then set up a full-blown firefly community science program. This involved regular firefly surveys on all trails of the forest where participants counted, measured, and tagged firefly larvae and females (initially, the male still had not yet been discovered). On weekends, data-gathering sessions were launched during which community scientists would record basic ecological data such as the depth of the leaf litter layer, the density of the forest canopy, and the humidity, temperature, and pH of the soil. Nearly a hundred local residents aged between five and seventy years signed up. Some participated for just one or two sessions, but many got so fired up that they became fully trained "firefly ambassadors." Two of them even published a scientific paper in the *Journal of Asia-Pacific Entomology* on how streetlamps along the paved paths at Bukit Kiara affect the firefly larvae (it turns out they do not like it).

One of the community scientists, Tan Boon Hua, recalls some of the nocturnal science projects: "We marked the larvae with a marker pen, and then doubled back over the same trail to see which larvae we have seen before. This data is then fed into a formula to estimate the population size of the firefly. This went on quite late into the night; we thought that the kids would not be able to take it, but no, they were running around spotting, here's one! there's another one!" With so much dedicated focus on such an obscure organism, many previously unknown details about its way of life and behavior were revealed by the community scientists. They saw fireflies feeding on snails (Tan: "That was really cool as well!") and a glowing female firefly guarding its nest of eggs in the soil. On another occasion, a message on the WhatsApp group announced that a hiker had finally spotted the elusive male of *Lamprigera*, so ad hoc night surveys were started to try to find more males (since finding out the exact species would largely depend on how the male looked). Tan: "We were all excited: Let's try and catch the male . . . *sample it*, I mean—I beg your pardon." In conclusion, Tan says, "Will I do this again? You bet! I'll do this again any time!"

The *Lamprigera* firefly (the group has not yet been able to figure out exactly which species it is, but it could well be a yet unnamed one) turns out to have been a golden choice as an emblem. Since that overwhelming evening in 2017, firefly events keep drawing huge crowds. "Anything to do with fireflies, there would be like . . . hundreds! It became obvious that there was a real magic to fireflies, there was a real pull to it," says Kribanandan Gurusamy Naidu, FoBK's president. Goh says, "Having a mascot, a big firefly, is something that would help a lot in protecting this forest. Yeah, I think it's really important to have a flagship species. People protect things that they know: in Malay we say, *tak kenal maka tak cinta*, what you don't know you won't love."

THE FOUR C'S OF URBAN CONSERVATION

And that is exactly how Michael Rosenzweig, an ecologist and evolutionary biologist at the University of Arizona, feels about it. Rosenzweig is not as well known among the general public as he should be. To me, he is one of the most inspiring ecologists, a clear thinker whose devotion to science and the natural world shines warmly through his lucid and concisely written texts. I remember spending whole afternoons excitedly devouring his 1995 book, *Species Diversity in Space and Time,* on the merbau floor of my house in Borneo at the time that I was working on my own book, *The Loom of Life*. In his later book, *Win-Win Ecology,* which is about reconciliation ecology (conservation in heavily human-dominated environments), he writes, "I know of no other branch of biological science that so involves lay people in its front lines. . . . However, reconciliation ecology is neither a religion nor a political philosophy. If it were, I would preach it. But, at its core, it is a branch of science."

I'd like to marry that quotation with what Karen Wong, one of the firefly ambassadors of Bukit Kiara described in the previous section, says in a testimonial produced by FoBK. Filming herself on her mobile phone from her own home (a clothes rack with laundry behind her), she looks straight into the camera, smiles, and states, "You do not need a huge background. But what you do need is a kind of curiosity; and also a sense of community."

Commitment. Community. Curiosity. Three key words. And courage: I would like to add courage to that list. Let us call them the four C's of urban community conservation.

Commitment. If you want to conserve nature in your neighborhood, you need to be in it for the long haul. It takes a lot of time and effort to get a baseline of the biodiversity and the ecosystem. And once you have that baseline, you need to keep monitoring, for a few good reasons: first, to record changes in the biodiversity (the comings and goings of species with time and with the seasons, the long-term declines or increases in species), and to record reshufflings of the ecosystem (which insects feed on that newly arrived exotic herb, how do mushrooms colonize the trunk of the recently fallen oak tree). But also to maintain a constant physical research presence, as they do in Validebağ, as a system to detect and deter encroachment.

Community. Urban nature conservation can never be a one-person job. Sure, a single enthusiastic, charismatic, well-connected person (like Rudolf Tenzer, mentioned in the preface) could be the initiator, nucleus, or motivator for a conservation project, but an entire community is needed for it to be a success. To begin with, even a small neighborhood greenspace can easily harbor thousands of species, and this huge biodiversity needs to be parceled out to a roster of community scientists to make the task manageable. But more important, only when an entire neighborhood is involved, either directly or indirectly, will a sense of shared responsibility, ownership, and guardianship evolve.

Curiosity. An urban greenspace is not just a lump of vegetation. It is a dynamic ecosystem with countless interconnected elements, full of fascinating natural history to be discovered, understood, and retold. Like being curious about the inner workings of the people you care for, community scientists are intrigued by the clockwork that makes their beloved local ecosystem tick. To gain an understanding, they will need to channel their curiosity in scientific ways: by making and recording observations, being systematic and repetitive about it, and by designing tests, drawing parallels, zooming in and out again to see the bigger picture, and all the other intellectual games that make science so much fun. The work of litter picker Graham Moates, who discovered the evolutionary trap of discarded bottles and cans in his neighborhood, stands as an example.

Courage. Not everybody is going to like what you do. In your own neighborhood, perhaps even in your own family, there may be people who think that urban paradise you are studying and protecting is just a vacant lot of weeds and rats better filled with houses for their children to live in. There may be project developers who nefariously question your findings, your legitimacy, and your motivations. Local authorities, too, may view your project with mixed feelings. They might support it (because it makes them look good), they might ignore it (because they have other plans), or they might do both, depending on the local politics of the moment. You are going to need the courage to keep working at it, trust your science, and keep repeating your beliefs based on the growing amount of information you are gathering.

And these four C's are not independent, of course. As the curiosity-driven scientific knowledge in your community increases, so will the love the community feels for the greenspace (*tak kenal maka tak cinta*). As your love increases, so will your commitment and courage. And as the size of the community of committed citizen scientists grows, the greenspace will increasingly become a commons for everyone to feel shared ownership of.

The examples from Turkey and Malaysia in this chapter (and from many other places described in this book) are successful, mature urban community conservation projects, driven by citizen science. But it can be hard to get such a project off the ground in the first place. If there is no prominent local naturalist or nature study group, the local community, even if it has the will to organize, may not have the time and scientific wherewithal to know how to begin, and initiatives may easily fizzle out, leaving the urban greenspace unstudied, unprotected, and ultimately lost.

For such situations, community science NGOs are springing up here and there. In KL, the Urban Biodiversity Initiative of Thary Goh helped the community around Bukit Kiara set up a firefly-focused community science program. In the Netherlands, my friends Aglaia Bouma and Norbert Peeters and I set up the Taxon Foundation, a network of biodiversity specialists with mobile labs that helps community groups draft a research plan and gets them started by doing initial assessments in mixed groups of laypeople and professionals and in makeshift labs set up in living rooms and garden sheds. We also help them get their stories out to the local and national media, and, once a self-supporting kernel of local community scientists is firmly established, we step back a bit but remain available with an online help desk for ecology- and biodiversity-related questions, identification requests, and moral support. And, on a small scale, it works. In the few years that Taxon Foundation has now been in existence, we have had a few successes and have kick-started a few community nature study groups that are thriving and growing.

We have known failures too. In one case, an old, insect-rich orchard marooned in the middle of a town near Utrecht was slated to be leveled for a new residential area. There was very little time, so we helped the neighborhood people do a midwinter biodiversity assessment. We formed teams and sieved hibernating insects from the piles of dead leaves and grass in the orchard and from behind the pear trees' loose bark. We dug up stones and lumps of clay to sample the mites and millipedes living in the soil. In two sheds in the neighborhood we helped the community scientists identify over two hundred species, some of them rare or new to the province or even the country. We arranged for local and national radio and newspapers to do news items. All to no avail. The report with the full biodiversity is still sitting on our website (and, presumably, in a drawer in the town hall), but the animals and the trees themselves are gone, buried under two streets of new houses.

When faced with such willful, shortsighted destruction of nature, with the intentional eradication by simple-minded powers of all the beautiful animals we painstakingly tabulated and illustrated, my old anger can sometimes resurface. Again I feel like that powerless, intimidated, but secretly proud boy being told off by the police inspector with the eyelids.

But that police officer is long dead, the police station where Isaac and I were held has now been chopped up and sold as apartments, and the soccer field has been replaced by a modern residential area with big gardens. When I visit my mother, I sometimes take a stroll there. The ditch we were defending still exists, but it is no longer polluted. It is wider now, with clean water, shallow banks, fringes of reeds, and nesting warblers, grebes, and waterhens. Dragonflies zip to and fro, a variety of water plants grows in the clear water, and water striders crisscross the surface. In other words, the ditch was restored to—even beyond—its former glory. Not because of our juvenile guerrilla activism but as part of the nature-inclusive design and landscaping that the residential area underwent when it was built five years ago. In the final chapter we will take a closer look at incorporating space for urban nature in city planning and architecture, a development that, under certain conditions, may be even more effective at urban nature conservation than activism.

19

LET IT GROW

I am surrounded by noise. Below me, the roar of the six-lane Kuala Lumpur ring road; above me, the shrieks and clanks from the construction cranes building a new apartment block; behind me, the hubbub from the throngs of the faithful who climb the 272 steps to the temples to pray. And on my right, a group of boisterous tourists taking selfies with the gigantic golden forty-three-meter-high statue of a Hindu deity. I am trying not to let the din distract me as I stand, precariously balanced, on a steep shaded slope, inspecting the vegetation growing against a cliff of the Batu Caves limestone hill.

I first climbed this one-square-kilometer hill, riddled with caves and covered in lush tropical vegetation, in 1989. At the time, it lay beyond the fringe of the city. Today the hill has been engulfed by it like a kīpuka by a Hawaii lava stream. It is a lumpy green emerald embedded in the gray Greater Kuala Lumpur conglomeration. I have veered off the main path leading up to the Hindu and Buddhist temples that have been built inside the caves to inspect the rich plant life of the hill slope. A vertical, shaded limestone cliff, moistened by a dripping mossy overhang, offers a botanical treasure trove.

The dense mix of leaf shapes that presents itself completely obscures the rocky substratum on which these plants have taken root. The feathery maidenhair fern *Adiantum malesianum* brushes against the large diamond-shaped leaves of the balsam *Impatiens ridleyi*. Here and there among them are the large, shiny, deeply veined leaves of *Monophyllaea hirticalyx*, a plant found exclusively on limestone and unique in the plant world for consisting only of a short stem adorned by that one single leaf; hence its common name, "one-leaf plant." In one place where the rock surface creates a small overhang and thereby remains relatively dry grow several stems of the shaggy-leaf fig, *Ficus villosa*, its mottled dark green, felty leaves, alternatingly thrown off the main stem to the left and right, hugging the rock surface closely.

I gently push aside the larger leaves of the *Monophyllaea* and the *Ficus* to reveal another layer of more modest inhabitants of the limestone cliff: a multicolored bed of mosses and lichens, with small mushrooms growing in between here and there. As I move my face closer to inspect these lesser occupants, I begin to notice more and more detail. Stocky brown millipedes with yellow spots move among the lichen. On a drier patch of rock sits a group of prettily coiled snails. A fat orange beetle trundles diagonally across the microvegetation, disturbing nimble crane flies that hang from the edges of the leaves of a small fern growing in between two differently tinted tufts of moss. Zooming in even further (my nose is now nearly touching the wall, and I allow myself to be fully absorbed in this mini-world), I observe how the vegetation seems to be responding to a fractal-like unevenness of the underlying limestone surface. Centuries of weathering have caused the rock surface to display that typical karstic appearance of pits within pits within pits. This has created microscopic as well as macroscopic variations in inclination, shadiness, and water retention, and each square centimeter offers slightly different conditions that particularly suit one specific set of vegetable occupants. The result is a rich quilt of different mixtures of tens, perhaps hundreds of species of fungi, mosses, lichens, algae, ferns, and flowering plants, with, of course, their own set of mini-animals attending them.

Behind this cliff that I am surveying, so abundantly clothed in tropical vegetation, lie the caves that now hold those Hindu and Buddhist temples but that, about three thousand years ago, were home to early cave-dwelling humans, whose cord-marked pottery has been excavated here by Malaysian archaeologists. Today those cave dwellers are long gone, but when I turn around and, from my high-elevation vantage point, look out over the vast conurbation of Kuala Lumpur, I cannot help thinking that what I am looking at is nothing more and nothing less than a jumble of millions of artificial caves, stacked one on top of another. When humans began building houses out of stone they essentially began recreating the caves that their ancestors had lived in: cool dry hollow spaces surrounded by stone walls, sometimes made from the very same material—cement produced from quarried limestone—that used to hold their ancestral cave homes.

And just as in those natural caves, wild vegetation will happily colonize the walls and roofs of our buildings. I descend the long staircase of Batu Caves, hail a taxicab on Jalan Perusahaan, and take a ride downtown, alighting in the city center. Avoiding the palm tree–lined main streets with their glitzy shop fronts, monorail stations, and parking lots, I venture into the back alleys, where the constant drip of condensation from air-conditioning units plays the same role as those dripping mossy overhangs at Batu Caves. Underneath each growling air conditioner I find formerly whitewashed walls, concrete ridges, and corrugated iron awnings supporting an abundant vegetation of green algae, black and orange fungus, invasive and native herbs, and even entire trees, obstinately rooting in cracks and crevices wherever they can. Some of the plant species are the same ones I saw growing at the caves earlier in the day. Others, such as a *Macaranga* tree, are opportunists that are the first to colonize any rooting spot as soon as it opens up. In between all this plant life I spot mosses, insects, spiders, and woodlice.

And yet, even though these grimy, untended back-alley walls, covered in slimy algae, unstoppable mosses, and unruly weeds, are undeniably "green walls" or "vegetated façades," when I look up those terms on the internet, I find images of completely different structures. According to Google's algorithms, a green wall or a vegetated façade is a section of outdoor or indoor wall, usually of a piece of hypermodern architecture, where a variety of plants have been artfully, sometimes geometrically, arranged in a more or less tightly packed pattern.

Superficially, these designer green walls, often a collection of dozens of different species of mosses, ferns, and herbs, resemble the natural vegetation that I saw up at Batu Caves. And, also superficially, they would seem to be a great way to help urban greening: a lot of urban biodiversity needs some form of vegetation, and by creating these aesthetically pleasing vertical parks on all those vacant walls (and roofs), we could give the urban ecosystem a much-needed boost.

I say "superficially" because a first difference from natural vertical vegetation would show up were we to look for anything else besides the green plants. Lichens, fungi, and small animals would be all but absent from these designer "living walls," at least initially. And if we dug deeper, we would not find a rocky support for this vegetation but instead a hydroponic mat, a steel frame, an impermeable base mat, and an intricate watering system of pipes and computer-controlled electric pumps to keep the green wall wet and alive.

Even that expensive system does not always prevent the green wall from ecological collapse. Enter the word dead as the first word of your search string and up come thousands of images of green walls gone grim and withered. The electricity supply of the watering system failed, the pipes got clogged, or the system was set too high or too low. An improper substratum was used, or the plants chosen required less sunlight, more water, a different pH, or all of those. Apparently, those lush vegetations that grow spontaneously in Kuala Lumpur's back alleys and on the cliffs at Batu Caves are not that easy to recreate—not even by green-wall design companies that unblinkingly will charge hundreds of thousands of dollars for installing one.

Why would this be? How could a ledge, accidentally placed in a moisture capturing position on a blank concrete wall, over time succeed in creating a perfectly functioning vertical ecosystem, where a carefully designed and constructed support system, planted with a deftly selected mix of plant species, would fail? One way to answer this question is first to have a better look at *natural* green walls. A scientist who has done this extensively is Rob Francis, a plant ecologist at King's College, London. I meet Francis at a conference at the University of Cambridge. He is a cheerful, unassuming academic with that embracing attitude that is nearly a given in field biologists who literally spend much of their time down to earth.

In 2011, in the journal *Progress in Physical Geography*, Francis published an overview of the types of spontaneous green walls you may find in a city. "To the casual observer, they are an unlikely source of ecological interest," he starts. But then he goes on to show how endlessly fascinating walls have been, and still are, to the ecologist. To begin with, European botanists since the seventeenth century have been eagerly documenting the plants growing on walls all over the continent. In an overview published in 1969, the botanical constituents of no fewer than 1,200 green walls are listed, ranging from the Roman Colosseum in Italy prior to its restoration in 1870 (when all plants were removed) to the walls of the Poitiers Cathedral in France. One of the reasons for this historical popularity, says Francis, is that walls often sport rare plants of rocky environments normally not encountered in the flat areas where cities are built. So, to city-dwelling botanists, those miniature rock gardens sprouting naturally on church walls might be an easy way for them to come into contact with true montane flora without needing to venture on far-away, expensive, and cumbersome field trips.

Those early studies, says Francis, often were little more than tabulations of the species found on walls. But since the 1970s, wall ecology has really come into its own, with researchers minutely dissecting the various factors that determine which plant grows where and when.

Reading his paper, and hearing Francis talk about it, you begin to realize that there is much more to wall ecology than meets the eye, and that the answer to the success of the Kuala Lumpur back-alley wall versus the failure of expensive designer green façades lies therein. To begin with, like any ecosystem, a vegetated wall does not come into being in one go but develops gradually over time. A newly built wall is first colonized by bacteria and algae, which always float in the air and can settle quickly and tightly in microscopic surface irregularities. The carbon dioxide that these organisms produce dissolves in the drip of rainwater, which becomes slightly acidic and starts weathering away the rock or stone of the wall as it runs down, creating larger pits that can provide a holdfast for mosses and small weeds. Once lichens and mosses begin to grow, and further weathering causes pits and small crevices to appear, the scene is set for the wall to begin trapping its own soil. Decaying organic materials from the microvegetation, combined with the fine dust that settles out of the air, create miniature soils that accumulate wherever there is an unevenness in the wall, such as a pit caused by an air bubble in concrete, a layer of mortar, a hole in a brick, or the upper edge of a lichen or a cushion of moss. Seeds blown in by the wind are washed down by rainwater and settle in one of these soil-filled pits. Their roots enlarge the pits, perhaps even forming a crevice in the cement.

As these crevices get larger and more numerous, and as the growing vegetation of herbs sheds more leaves, the remains of which provide further nutrients that partly stay behind in these pits and crevices, a new stage of wall ecosystem development starts, with larger plants settling and enough living and dead organic matter accumulating for small invertebrate animals to creep up. Woodlice, springtails, and snails begin exploring the wall and, when they encounter pockets that are moist and nutritious, lay their eggs there. Ground beetles and centipedes begin feeding on these developing populations of detritus-feeders, and once the populations of these predators reach sufficient numbers, perhaps even a lizard might take up residence on the wall, feeding on the abundant insect life. The lizard droppings and the continuously growing, diversifying, and ever-deeper-rooting vegetation begin to create decently sized pockets of soil that stay moist even during longer periods without any rain, allowing moisture-loving plants to join the rock plants already growing there.

I could go on, but I think I have made my point: the outside of a building, from the moment it is put up, is like a mini-Surtsey, the pristine volcanic island risen from the seafloor that we saw in an earlier chapter, ready to be slowly colonized by whatever species the winds and the rains bring in, and, slowly and erratically but surely, mature into a true ecosystem. The key point is that nobody, no human designer or landscape architect, selects which species go where. It is the wall itself, its physical structure and the alkalinity of its materials, and especially the random and contingent process of ecological development, that determines the eventual composition of its flora and fauna—a composition, moreover, that is never constant. A wall is a vertical island, and as island biogeography teaches us, even when eventually a sort of equilibrium is reached, species will keep coming and going.

A BLANK CANVAS

How different it is with those human-designed green walls! Instead of letting nature take its course, designers sitting at their desks decide on selections of plant species picked out from the catalogues of garden centers. These are then stuck into a uniform substrate, and hey, presto! You have a living wall. But from the moment the irrigation system is switched on, the ribbons are cut, the invoice is paid, and the building is officially opened, the green wall might be dying a slow but sure death. The awkward starting point—a mature herb vegetation composed of species that were forced to live together—and the absence of entire echelons of the ecosystem, for no mosses, algae, bacteria, lichens, invertebrate animals, or fungi were included in the mix, together are a recipe for ecological collapse.

Why would we choose to create such poor (but expensive) imitations of natural ecosystems if nature can do this for us better and for free? Why do we not simply offer a wall, preferably one that is prepped for quick and successful settlement by flora and fauna, and let the process of ecosystem assembly take its natural course? Why, in other words, do we not let nature do the selecting of the species allowed to live there, rather than a human designer?

First, of course, is the fact that many walls that are used for construction have limited possibilities for supporting vegetation. Their pH tends to be very high (very alkaline, that is) and they are too hard to allow fast weathering, insufficiently porous to absorb a lot of moisture, and too thin to maintain a stable temperature and humidity. In other words, we should develop building materials that maximize the suitability for plant growth while at the same time maintaining structural integrity. Today this is not too much to ask. In the old days, if there were trees growing from your house and moss on your roof, it was considered an ominous sign that your house was about to collapse. These days, we have the technology to develop smart materials that can do both, become vegetated while at the same time supporting the building.

But there are two other reasons why we prefer to design green walls rather than let them grow spontaneously. These reasons are purely human and therefore much harder to deal with. First, natural ecosystem assembly takes time. For a blank wall to become covered with a climax vegetation, a kind of stable end state in the succession of becoming vegetated—or at least the greatest climax possible, given its materials and location—can easily take years if not decades. And we usually do not have that sort of patience; we want our urban ecosystem now! So we decide to lend nature a hand.

The second reason is that we have preconceived ideas (enhanced by the images in coffee-table books on green design) about how such a green wall should look. What we imagine is something similar to that lush, moist vegetation of broad-leaved herbs and ferns that I witnessed at Batu Caves, perhaps interspersed with some soft emerald cushions of moss. But that kind of vegetation is only possible in extremely humid, warm locations, such as a shaded limestone cliff in the tropical rainforest or the sides of a gorge with a waterfall somewhere in the wet northeast of Canada. The lush vegetation that I found in that back alley in Kuala Lumpur notwithstanding, what you most likely get if you were to let a wall in a city become vegetated spontaneously would be an irregular mixture of grasses, spindly roadside weeds, stinging nettles, and brambles, half-eaten by weevils and slugs. The wall material would remain visible here and there, and for the rest it would be largely vegetated by "undesirable" invasive plants. During dry spells, it would go brown and yellow for weeks or months, its shed dead leaves piling up at the foot of the wall. In other words, it would be what biologists may call a dynamic ecosystem but what other people call an eyesore.

What we have here, then, is what is known as a wicked problem—one that needs multiple, potentially contradictory actions to solve. First, of course, there is an economic counterforce. For the booming world of green wall designers and the companies and authorities that hire them, it is important to maintain the idea that successful green walls are expensive, high-tech, and need to be designed, built, and maintained by professionals. But an even greater problem lies in our mindset. We have deep-rooted, long-held, and pervasive ideas about the amount of freedom that we allow the nonhuman parts of our surroundings to have. We need, in other words, to look at the world in a different way.

One group of people whose bread-and-butter is to help us change our perspective are artists. All over the world, bioartists and land artists, but also gardeners and landscape architects, are exploring how humans influence the landscape and how we view the freedom (or lack thereof) of our nonhuman coinhabitants. The Austrian artist Lois Weinberger, for example, created the artwork *Wild Cube*, a cage that looks like a prison, installed in the urban environment. But it is an inverted cage: rather than keeping humans in, we are kept out so that wild nature can live inside without any human intervention. Weinberger began creating these Wild Cubes in the 1990s, and some of them, such as the twenty-meter-long cage he built in the Austrian city of Innsbruck, now contain complete caged forests. The Wild Cubes, and especially what happens at the interface between the space inside the cage and the space outside, where we are, drive home the notion of what wild nature in that particular place and time would be like if we were not constantly hammering it down.

Another exponent of this school is the French gardener Gilles Clément. In Lille, in 1995, he created *L'île Derborence*. On a vast lawn in the city's Parc Matisse sits an irregularly shaped island of 0.3 hectares, raised seven meters above the ground by organically shaped walls made from the rubble of the buildings that were demolished to make space for the park. Growing on top of this inaccessible island is a completely wild and spontaneous vegetation. Again, the island, a wild enclave in an otherwise tidy urban park, poses the question of how we feel about allowing a bit of wilderness in our organized, safe, human world. One answer to that question is the fact that, recently, and much to Clément's chagrin, the municipality has planted a row of trees in front of the island to reduce the confrontationality and make it blend in more with the rest of the park.

It is surprising that the green-wall concept so far seems to have resisted any interest from this artistic movement. Perhaps this is because the world of green walls is already well established, and artists may feel not much new can be done with them that has not already been done by the established green-wall construction companies.

I think this is a missed opportunity. There is great potential for artists to work with community scientists on the transformation of blank walls—canvases, if you will—into vertical ecosystems. "Work" here would not be about planting stuff because nature will do the planting itself in its usual haphazardly expert fashion. Instead, the group could create a substrate that resembles the natural rock surfaces that are so abundantly vegetated in nature, perhaps even a surface that is interesting and attractive to look at before it becomes vegetated. Better still: by creating small-scale variations in acidity, porosity, water-retention capability, et cetera, it might even be possible to slowly let a predefined pattern or an image appear on the wall once different plants start colonizing it. The group could also set up a longlasting biodiversity science project to see which species come and go and how the neighborhood wall ecosystem embarks on its own idiosyncratic, tortuous route toward a biodiverse vertical urban wild space. Like parents raising an unruly child, a community that nurtures a green-wall ecosystem from birth to maturity will not mind it being a little messy and untidy.

WEAPONS OF MESS DESTRUCTION

In fact, we need to view terms like "messy" and "untidy" with more than a little suspicion when used in the context of urban greenspaces. "Eyesore" is another one. "Weeds" and "unkempt" should be flagged too. One person who put his contempt for such terminology and the sentiments behind them into words was the Welsh entomologist Douglas Boyes. In his tragically short life, Boyes made a name for himself as a UK moth and butterfly specialist (he started studying them at age twelve), a data collector on local biodiversity, and a PhD student at Newcastle University working on the impact of artificial light on moth populations. Scientifically, he made a splash with a paper in *Science Advances* on August 25, 2021 (just eleven days before, at age twenty-five, he took his own life—he suffered increasingly from depression), in which he showed that modern LED street lighting is actually more damaging to insects than the old-fashioned yellow sodium kind. But perhaps of even greater impact was an essay he wrote in 2018, when he was only twenty-two years old, on his personal blog, douglasboyes.co.uk, titled, "Obsessive Tidiness Disorder, or: How We Can Learn to Stop Worrying and Love Nature's Messiness."

As an entomologist, Boyes was keenly aware that nature is essentially messy, rambunctious, unruly, and unkempt, and that biodiversity blossoms only where nature's obstinacy is given free rein. It is humans' obsessive tidiness that agitates against the rambunctiousness of nature, and we do so with "weapons of mess destruction," as Boyes called them. Think of high-pressure washers to remove from pavements the complex microbial ecosystems of bacteria, fungi, algae, and mosses, like the ones we admired in a previous chapter in Paris; the leaf blowers wielded by gardeners, who annually send tens of millions of tons of bagged dead leaves to landfills; the lawnmowers and garden shears for combating plant vegetation—or worse: hot steam, flames, herbicides, and pesticides. Boyes wrote, "This universal expectation of tidiness has been aided and abetted by increasingly powerful machinery to mow, trim, strim, saw, blow and flail. The casual destruction of habitats has never been so easy." And as people wielding these weapons convert nature's messiness into an imposed tidiness, insects and other animals make a hasty retreat.

In his beautifully crafted text, Boyes makes the point that we have to learn to see our surroundings with new eyes. We have to think like an insect, a snail, or a millipede, not like a human burdened by centuries of preconceptions. Tidiness is a cultural concept, and cultural concepts can change. Start letting a corner of your garden run wild, and proudly explain the how and why to your neighbors, he advises. Local authorities, too, should make it clear to complaining unenlightened residents that they sometimes leave grass uncut and leaves unblown for solid ecological reasons. As Boyes says, "Where it causes no harm, why can't unkempt nature be the norm? . . . If we want to save our wildlife, we must learn to love nature in all its messy glory."

One way for budding urban naturalists to begin this love affair is by exploring their city for wild, unkempt places that have developed into habitats under their own steam, without any human intervention. These need not necessarily be large urban forests or expansive overgrown vacant lots; they can be the tiniest forgotten corners. In fact, the smaller the better, because the greater the chance that they will evade detection by the tidiness police. Think of those spontaneously vegetated façades described at the beginning of this chapter, but there are also less obvious spots.

Jonas Frei, a young Swiss urban botanist, illustrator, and author of the book *Stadtwildpflanzen* (City Wildflowers), has made it his mission to find wild mini-ecosystems where you would least expect them. You can find him (in man bun, goatee, long coat, and jeans) on his hands and knees on a Zurich pavement, examining, sketching, and photographing miniature jungles of wild herbs. A lush mix of dandelion (*Taraxacum officinale*), field bindweed (*Convolvulus arvensis*), goosefoot (*Chenopodium*), and several moss species huddling among the cobblestones underneath some parked bicycles, . . . A flurry of *Onoclea sensibilis* and other ferns peeping through the slits in a manhole cover. . . . An awkwardly positioned (and therefore untended) nook in the wall of a house shared between young loquat (*Eriobotrya japonica*), male fern (*Dryopteris filix-mas*), and ash seedlings (*Fraxinus excelsior*). He even manages to document how these mini-communities change over time: on the Instagram page that he made for his book is a time-lapse movie of the comings and goings of several species of grasses and herbs in a pothole in the tarmac of a Zurich road over a period of five months.

In Amsterdam, his Dutch colleague Ton Denters is often seen in similarly compromising positions: nose to the ground, camera and notebook at hand, exploring microvegetation under bridges, in the nutrient-rich seepage around public toilets, on top of canal walls in the Red Light District, and in grimy drains. In August 2021, on one of his botanical excursions through the area directly east of the Central Station, he came across an artificial lawn. Made from artfully interwoven green plastic strands, this low-maintenance substrate is increasingly replacing real grass lawns all over the world. Where manicured grass lawns already have a minuscule biodiversity, you might expect that such monofilament ribbon pile product, as it was originally called (better known nowadays under the generic name "astroturf") is even worse, with zero biodiversity.

But what Denters noticed when he subjected the stretch of astroturf to close botanical inspection was that real living plants were sprouting everywhere in between the artificial blades of grass. He took out his tape measure and demarcated a standard ten-by-ten-meter botanical plot. Then he began identifying the living plants he found growing among the astroturf. He found some very ubiquitous weeds, such as dandelion (*Taraxacum officinale*), chickweed (*Stellaria media*), and daisy (*Bellis perennis*), but also some seedlings of trees such as elm (*Ulmus*) and white willow (*Salix albus*), and a whole bunch of other weeds: marsh yellow cress (*Rorippa palustris*), toad rush (*Juncus bufonius*), horseweed (*Conyza canadensis*), hoary plantain (*Plantago intermedia*), and many more. All in all, in what turned out to have been the first standardized botanical inventory of a piece of astroturf ever, he recorded twenty-five different plant species in this patch of make-believe habitat. That is twice the number of species you would normally find in a real grass lawn of the same size!

Why? Part of the answer lies in a 2004 paper published in the journal *Functional Ecology* and titled "Astroturf Seed Traps for Studying Hydrochory." As it turns out, the labyrinthine structure of astroturf is excellent for trapping seeds that are dispersed in flooding water (this is what "hydrochory" means), and probably the same is true for seeds and other particles that spread in the air. The slow accumulation of grains of dust in between the plastic ribbons eventually creates a soil that holds water, and the seeds that are caught in the green maze grow up under what must be ideal conditions: humid, protected from direct sun and the wind, and, most important, shielded from competition with other plants. Eventually the plastic lawn that Denters studied could become a vegetation with a remarkably high biodiversity. In fact, in some places in the UK this is already happening, as is evidenced by sightings of greens maintenance workers doggedly mowing astroturf lawns.

ACCIDENTAL JUNGLES

Of course, it would be better if artificial grass did not exist in the first place, for a multitude of reasons. But while it is on the increase (the global fake grass business is growing by some 15 percent each year), we had better be informed about its role in generating spontaneous vegetation. So it would be wonderful if community nature study groups took it upon themselves to start monitoring the biodiversity of artificial lawns.

Many community science programs that embrace spontaneous vegetation in cities already exist. The fun tradition of using colorful chalk to jot down the local and scientific names of wild plants growing in your neighborhood streets, originally dreamed up in 2014 by French artist and comedienne Frédérique Soulard, who named her project *Belles de Bitume* (Tarmac Beauties), has now been adopted by individuals and groups in cities in many other countries. In Frankfurt, Germany, it is known as *Krautschau* ("herb show"); in Nottingham, UK, as *wildflower graffiti*; and in Leiden, the Netherlands, as *botanisch stoepkrijten* ("botanical pavement chalking").

The Dutch artist Hans van Lunteren found a different way to give a voice to spontaneous city street vegetation. For his project *Toevallig groen* (Accidental Green), people can register saplings of trees that have spontaneously sprouted somewhere in their neighborhood. So far, hundreds of cherished trees from cities in the Netherlands and Germany have been registered and placed on the project's interactive map—especially in Utrecht, where Van Lunteren is based, and where the municipality provides plastic labels that the saplings can be tagged with to signal the greens workers that these "illegal settlers" should stay.

A woman named Liesbeth, for example, tagged and uploaded a picture of a young rowan (*Sorbus aucuparia*) emerging from the bicycle rack in her street. “It sprouted spontaneously two years ago,” she writes. “We are happy with it.” And Lot van Hooijdonk adopted a fig tree (*Ficus carica*) growing from one of the old walls of the canals in Utrecht’s medieval city center. Until recently, Holland was too cold for this Mediterranean species to survive, but climate change and the urban heat island have conspired to permit more and more figs to survive in the urban wild. This particular one, though, sprouted near an exposed central heating pipe, providing just the right balmy microclimate for its roots.

I love this project and I hope it will be adopted by community groups elsewhere in the world because trees in particular are elements of the urban ecosystem for which people mistakenly think they require planning and maintenance. Tree planting projects in cities, laudable as they are, give the false impression that trees need assistance to take root, that they cannot plant themselves. They can, they do, they will, and, I think, they should. Like all plants, trees scatter their seeds far and wide, aided by wind, water, and animals, and their seedlings will sprout in a million places. Many will be dead on arrival because they grow up in a microhabitat that is outside the tolerances of this tree species: too warm, too cold, too wet, too dry, too acidic, too salty, too trodden. But some will land in a site that is exactly right and grow, first into saplings, then into mature trees. It is these saplings that have successfully planted themselves that we should rightfully cherish in the way that *Toevallig groen* encourages.

Imagine a city where trees, and ideally all plants, could largely be allowed to settle spontaneously. The air and water of the city are loaded with seeds and spores of plants that are in the process of evolving into perfectly adapted urban plants. Allowing them to colonize their own niches wherever they find them in the city, modified and curbed, of course, by human needs and limitations, would result in biodiverse, naturally assembled ecosystems, instead of the artificial species combinations that largely depend on the whims of human public greens managers.

One way to stimulate this would be to invent, develop, and use building materials that encourage the natural development of ecosystems. At the beginning of this chapter we had a look at this for vertical vegetation, but on horizontal surfaces it is possible too. The Dutch company

Gewildgroei, for example, created "living pavement" tiles with holes for wildflowers to grow in: you still get a functioning pavement, but also an ecosystem. Incidentally, *gewildgroei* is a pun on the Dutch word *wildgroei*, which, with a strongly negative connotation, means the wild, uncontrollable growth (of anything). The newly invented term *gewildgroei*, on the other hand, means "desired growth." The company's name has become a branded trademark for any kind of welcome spontaneous vegetation, and in 2022 even made it into the newest edition of the authoritative Dutch *Van Dale* dictionary.

Some botanic gardens are even beginning to include spontaneous urban vegetation as "habitats" in their own right in their displays. They will reconstruct parts of city pavement and show the weeds growing in the cracks between the pavement tiles, thereby raising the botanical status of this type of "trashy" green.

The Polish artist Diana Lelonek collects discarded human-made objects (a high-heeled shoe, a broken plastic bottle, a piece of computer hardware) overgrown with mosses and other plants, and some of her finds are continuing their lives as minivegetations in the botanic garden of Poznań. The botanic garden allowed her first to use an abandoned greenhouse in the back of the garden. "The gardeners started to care for the garbage plants," she says. "These rubbish-plant hybrids started to be treated the same way as other plants there." Eventually she persuaded the management of the garden to move her objects to the part of the garden that is open to the public. "I made cards exactly as the ones they had in the rest of the garden. . . . The Botanical Garden in Poznań has a very traditional approach to plants, so it was difficult to explain to them that this wasn't just an artistic project. I was also looking for new classifications for these garbage-plants. So, I started to call them post-Adidas habitats, polymer habitats."

Polymer habitats. It is just one of the many words we have come up with to describe this new urbanized reality of ours, a world in which wild organisms colonize and appropriate the artificial components that humans have inserted into their world. A set of habitats that is unlike any that has ever existed in the history of life on Earth and that therefore deserves to be explored and studied by us as if our lives depended on it.

CLOSER

Readers tend to underestimate how much writers enjoy it when their readers get in touch with them. Sometimes a reader writes to me in a prose that is much more beautiful than what she compliments me with. Neha Dsouza from Mumbai, India, who allowed me to quote her, writes that she is "by no means a naturalist." She goes on: "I have a boring corporate job that doesn't afford me the opportunity to pause and take notice of the world around me. Through your book, however, I found a gateway into a world, that I had hitherto not taken notice of. I live in Mumbai, and as you probably know, it is one of the most congested and crowded cities in the world. And yet, after reading your book, I started paying close attention to the trees, plants, birds, and flowers in my locality and a bittersweet feeling settled upon me as I realized that I have lived twenty-eight years of my life without noticing the wonders right outside my window."

Though she claims to be no naturalist, as her words continue, it shows that she has become one. She writes, "I was especially impressed with the Indian house crow and the Indian pariah dog, and how they have adapted to our concrete jungle. Right outside my window is a lush tree, that on further investigation, I recently discovered is the *Calliandra* or 'Powder puff.' A quick search on the internet revealed that this tree is common in Latin America. I began to imagine this tree's backstory and the journey it may have taken to find its roots in Mumbai, a city so far away from its place of origin. Similarly, I discovered several *Mussaenda philippica* trees and *Lantana camara* shrubs growing all over my locality. It amazes me how these plants have traveled across countries and sometimes continents and managed to thrive on our concrete pavements. In addition to making me marvel at the way nature thrives and adapts to new environs, it also made me contemplate how nature has no boundaries, is confined to no borders, needs no visas, and doesn't ask foreign plants to go back to where they belong. There is so much to learn and emulate from nature."

Although Dsouza is referring to one of my previous books, it is people like her that I had in mind when I began writing *The Urban Naturalist*: humans from all walks of life whose eyes may be opened to the nonhuman community that they are part of, and all the peculiar ways in which it is shaped and molded by our activities. Perhaps Dsouza will continue her botanical explorations of her neighborhood, perhaps she will become the go-to person for matters concerning local flora. On her way to work, she might begin investigating the animals that feed on the exotic trees that line the roads. She might even become an activist, campaigning for the survival of this roadside tree or that neighborhood park, or a bit of forgotten swampland by the railroad. Then again, she may be content simply to continue marveling at the animals and plants that she shares her street with and how they got there, and take comfort from having discovered so many new nonhuman neighbors.

Under some circumstances, nonhumans are the only creatures around that offer such comfort. In 1986, Kenyan pro-democracy activist Patrick Onyango, then a teacher, was held captive by the dictatorial regime of Daniel arap Moi. Incarcerated in the Naivasha prison and subjected to daily torture sessions in the infamous basement of the Nyanyo House government building in Nairobi, insects were his only friends. "I welcomed the company of the mosquito," he says, which helped him "to cope with the prolonged solitary confinement and remain in control of my faculties. Mosquitoes were able to get into my cell through the grilled windows from the outside, the free world, and keep me company for a bit, then fly away. I would send the mosquito to pass a message to my family members of the free. There was a particular mosquito, which could land on the same place of my right-hand wrist, fill up, and fly away. This ritual would be repeated many times."

There is comfort in the sensation of being part of the multispecies community that populates our cities, especially at times that our human surroundings feel oppressive, demanding, or hostile. Animals and plants are not resentful; they will tolerate and forgive us and continue to cohabit with us regardless of all the injustice we as a species inflict upon them. We should give them credit for that. At the very least, we should acknowledge their presence.

And, as Dsouza points out, this applies especially to species that have come from afar. Just like its human counterpart, the urban world of plants and animals is a melting pot of species from all over the world. Thrown together by our indiscriminate economic activities, and despite having no shared ecological background, they build new ecosystems against all odds and with quiet ecological determination. Rather than applying xenophobic sentiments to other species and vilifying plants and animals for nothing but their nonindigenous background, we should applaud them for adapting to each other and to the local species, and for helping to craft these novel ecosystems. Just as with human urban communities, we cannot do this without the immigrants, who bring the necessary flexibility and wherewithal to build functioning, dynamic ecosystems in the extreme environment of the city. I have said it before and I will say it again: in urban ecology, there is no place for the combat of exotic species. We need every species we can get.

There also is no place for valuing one species over another. It is the urban ecosystem as a whole we need to conserve, not necessarily the individual species in it. Ecologically speaking, orchids are equivalent to any other wildflower. We may have a cultural, aesthetic predilection for them, but do not let that cloud your ecological judgment. We may love ladybirds, hedgehogs, butterflies, and swifts, but in the ecosystem of the city, they are not more (or less) valuable than dandelions, mosquitoes, brown rats, cockroaches, and city pigeons.

I am also hoping this book will persuade you, reader, to take the plunge into community science. Make use of iNaturalist, Observation International, or any of the other biodiversity recording platforms to get in touch with kindred souls in your neighborhood. Join a community lab if there is one, or start one with friends if there isn't. Build your own home lab with whatever equipment you can rustle up. Start a collection and hone your skills and expertise.

And I hope I do not come across as patronizing from my privileged position at the academic high table. The son of a science enthusiast, and having started out as precocious naturalist schoolboy, I am a community scientist at heart. I may have had a career in universities and other official research institutions, but my soul has never left my childhood attic with its cigar boxes with pinned bugs, home-made bat detector, and chemicals from the local pharmacy.

During my time working in academia, I have seen how professional science has begun drifting away from that romantic ideal. When, still a student, I entered the office of my future MSc supervisor, I found him surrounded by his own private laboratory. He had a microscope on his desk, specimens in jars with alcohol in a rack by the wall, in the corner an aquarium where he was conducting experiments with freshwater snails, and many meters of books, old and modern, along the walls. Today, four decades of centralization and health and safety regulations have whittled away at this way of working, and I, holding the same position as he did then, command just a desktop computer and a desk in an open-plan office. The specimens are safely hidden away in the central collection facility. The books are held in the library, and experiments may only be done in the central lab facility after filling out the necessary forms.

I understand the rationale and the efficiency-driven motivation behind this, freeing up time for scientists to write publications, forge large international collaborations, and write grant applications, but it makes for a decidedly bland and unromantic version of science. I often liken scientists in such situations to William Pearl, the character in Roald Dahl's short story "William and Mary" whose brain and one eye were kept alive after his death. Scientists in many an academic institution are similarly stripped of all their peripheral facilities and reduced to merely a brain.

But science is so much more than simply a cerebral activity. To me, the kinesthetic pleasures of running after a bird and trying to fix it with one's binoculars, or carefully dissecting the genitalia of a slug, or gently pipetting the DNA-containing supernatant off the protein layer in a DNA extraction, or thinking up and building makeshift equipment, or catching a speedily flying jewel beetle from midair with an expertly launched flick of the net, or kneeling by a flower and sketching the shape of its petals, or agitating the leaf litter sieve with the exact right stir-frying movement, or selecting insect pins from a catalogue—all that and the skills one hones while doing any of these activities are what makes science fun and exciting.

In 2016, aged fifty and having reached that pinnacle of academic career achievement, a full professorship, and the leader of a research department, I decided to follow my heart, away from academia and into community science. Maintaining only a small part-time position at the museum and the university to continue to supervise my graduate students, I became a mostly independent scientist, seeking the general public rather than students and academic peers as my audience. I began giving lectures, writing books, and organizing expeditions for a public of people interested in the same scientific topics that drive me.

Iva and I set up the scientific travel agency, Taxon Expeditions, and organize real scientific expeditions to both urban and nonurban destinations for mixed teams of laypeople and international biodiversity experts. With my friends Aglaia Bouma and Norbert Peeters, I founded a non-profit, the Taxon Foundation, to assist neighborhood groups to study their own urban environments. We rented a space in a center for artists and startups in an old school building around the corner from our house where we are now building a community lab full of microscopes, DNA equipment, and a natural history collection for everybody to use, work with, and be inspired by.

And it is so much fun! Working with people whose dream it is to do real science in their own environment and who are thrilled by the privilege of bathing in the expert knowledge of scientists with a lifetime of experience has made me fall in love with science all over again.

Glowing with the refound enchantment with science, I visit the engraved stone in the garden of my parental home that marks the place where we scattered my father's ashes. In the thirty-five years that have passed since his departure, the stone has become part of the ecosystem. It is covered in algae and lichens, and woodlice live under it. Around the stone I spot many fragments of garden snail shells: a local song thrush has chosen it as the anvil on which to smash its prey.

ACKNOWLEDGMENTS

Large parts of this book were written during three writing retreats. First, in November–December 2021, at the Fundación Cultural Knecht-Drenth in the magical hilltop town of Callosa d'En Sarrià, Spain. I thank Jean-Marie van Staveren and Yolanda Blackstone for hosting me there. Next, in September 2022, at the unbelievably comfortable and welcoming Fondation Jan Michalski in Montricher, Switzerland, where Chantal Buffet and Guillaume Dollmann were my hosts; the incredible Vera Michalski-Hoffman is the one ultimately to thank for including me in the 2022 nature writing program. I had the privilege of living and working with a most inspiring and wonderful group of fellow writers in residence, namely, Katarzyna Boni, Florin Irimia, Safya Bakhtyari, Yvonne Adhiambo Owuor, Philippe Gerin, Toine Heijmans, and Katya Apekina. And finally, in April 2023, in the cozy, secluded Het Schrijvershuisje in Lutterzand, the Netherlands, of Dorine Holman.

As always, my agents, Peter Tallack of Curious Minds literary agency and Louisa Pritchard of LPA, were the ones who helped convert my book idea into a proposal that led to a deal with MIT Press, where Beth Clevenger was a marvel of an editor and coach, who made what could have been a bumpy ride into a smooth one. Deborah Cantor-Adams and Marjorie Pannell did a great editing job. My friends Frank van Rooij and Aglaia Bouma were again my faithful proofreaders, as were Iva Njunjić, Jonathan Silvertown, Marianne Pinckaers, César Coll Alfeirán, and Imke Smeets.

The collaboration with my friend, the illustrator Jono Nussbaum, was a delightful experience. I thank him for finding the exact right visual atmosphere to open each chapter with, for the good times we had in his house in Tuscany and in Castello di Potentino.

In writing this book, I benefited from the kindness and generosity of many colleagues and friends who donated their time. Cihan Babuccu showed me around his cherished Validebağ Grove and introduced me to its flora and fauna, and his efforts to conserve it. Emrah Çoraman toured Iva and me around his campus in Istanbul, treated us to Turkish coffee, and told us about his plans to make the university and the whole of Istanbul more biodiverse and more nature literate. During our "taxathon" in Noorderheide, and later in an online chat, Patje Debeuf told me how he had become a naturalist. In Belgrade, Jovana Bila Dubaić showed me and told me about the many feral honeybee colonies she had discovered, and Andrija Filipović explained to me how the media drive a view of exotic species as fearsome and loathsome. In an online interview, Thary Goh told me about his Urban Biodiversity Initiative and the use of the giant firefly as conservation mascot in Bukit Kiara forest, Kuala Lumpur. Sigrid Jakob shared her unexpected route from renegade to community scientist with me. In an online and a live meeting, Jan-Maarten Luursema gave me his view of tinkering and community labs. In a long series of emails and video files, Bram Koese told me everything about the guerrilla conservation project that he started to make commuters aware of the toll that roadkill takes along a road near his hometown. With Norbert Peeters I worked on a book chapter about Mary Treat, and he kindly allowed me to reuse and rework the material for this book. Andrew Quitmeyer was a generous and unbelievably fun host when I visited his lab in Singapore and has been helping me with lots of other details by email ever since. My mother, Lineke Schilthuizen-van den Berg, delved into her unparalleled, uncluttered memory and gave me an insight into our family background, remembering details of events that took place nearly a century ago. In Istanbul, I spoke at length with Rana Söylemez about the Roma Bostan urban farming project that she helped set up. Jasper van Kouwen took Iva and me on an urban speleology expedition of our hometown and showed us the hidden human-made underground world (and we showed him the underground ecosystem that exists there). Patrick Onyango gracefully shared with me the horrific experience of being held captive and tortured as a political prisoner in Kenya and how he drew comfort from being visited by a mosquito each night.

Many others—colleagues, friends, family, readers, even complete strangers—helped me with details, bits and snippets of information, and essential acts of kind assistance. These were Florian Altermatt, Katarzyna Boni, Nico de Both, Eli Broekhuis, Maxime Dahirel, Richard Delval, Lee Dugatkin, Neha Dsouza, Rob Francis, Barbara Gravendeel, Mike Groenhof, Auke-Florian Hiemstra, Steven IJland, Florin Irimia, Praveenraj Jayasimhan, Marc Johnson, Alex Kemman, Niels Kerstes, Sotiris Kountouras, Oleg Kovtun, Sophie Lokatis, Jolanda Maas, Juan Millan, Kees Moeliker, Aldo de Moor, Martha Moss, Yvonne Adhiambo Owuor, Marianne Pinckaers, Liselotte Rambonnet, Bruce Robertson, Lisa Scheifele, Fenna Schilthuizen, Jan Schilthuizen, Dan Stowell, Geert Timmermans, Nedim Tüzün, Robert Vargovitsh, Aaf Verkade, Kitty Vijverberg, Vincent Wittenberg, Wan Yusoh, the team of translators at the Arboriculture conference in Vitoria de Gasteiz, Basque Country, Spain, and the students of the Leiden University courses of Urban Studies and Urban Ecology & Evolution. Though their contributions may have been small, they were often crucial to overcome a dead point, and I am very grateful for that.

NOTES

INTRODUCTION

When I began writing this book, I came across a few other authors who had also written chapters or essays with a title and ideas similar to those in *The Urban Naturalist*. A chapter with the title "How to Be an Urban Naturalist" appears in Brenner, *Nature Obscura: A City's Hidden Natural World*, and an essay with a very similar title can be found in J. C. Smith, "How to Be an 'Urban Naturalist' (& Save the Planet)." There is also Garber's book, *The Urban Naturalist*. While I developed the title and subtitle of my book independently, I wish to acknowledge these sources as representatives of the same type of motivation.

CHAPTER 1: THE AGE OF THE AMATEUR

The text on Mary Treat is a loose restructuring and rewriting (in English) of the Dutch text by Schilthuizen and Peeters, "Mary Lua Adelia Treat," and was checked and approved by Norbert Peeters. I also used Peeters and Van Dijk, *Darwins engelen,* and Gianquitto, *Good Observers of Nature.* The quotations are mostly from Treat, *Home Studies in Nature*. The book in which Treat is fictionalized is Kingsolver, *Unsheltered*. The Darwin Correspondence Project can be found at https://www.darwinproject.ac.uk/letters. The text on Edgar Allan Poe and discussions of the amateur science movement in the United States are based on Tresch, *The Reason for the Darkness of the Night,* and the episode about his work on the *Science Friday* podcast of July 2, 2021. Information on Sachs denigrating the experimental work of the Darwins is from Kutschera and Briggs, "From Charles Darwin's Botanical Country-House Studies to Modern Plant Biology."

CHAPTER 2: ROCK VACATIONS

My mother filled in some gaps in my understanding of my father's youth and upbringing. The geological club magazine that my father wrote for is *Gea*, published by Stichting Geologische Activiteiten (www.gea-geologie.nl), and his articles that I mention are J. G. Schilthuizen, "Ponskaarten als Hulpmiddel bij het determineren van mineralen" (1978), "Zelfbouw van een geigerteller" (1979), "Determineren met boraxparels" (1979), "Steenzaag met vertikaal draaiend blad" (1981), and "Kristallen tekenen met uw computer" (1987). The Museum ten Bate van het Onderwijs (colloquially known as the Schoolmuseum), originally housed in Hemsterhuisstraat, is now called Museon, at the Stadhoudersiaan, The Hague (oneplanet.nl).

CHAPTER 3: DARWIN @HOME

The scene at the start of this chapter took place during the "taxathon" of Taxon Foundation at Noorderheide, June 14–17, 2022. My interview with Patje Debeuf was conducted by Facebook Messenger in February 2023. More background on the ObsIdentify app is in Schermer and Hogeweg, "Supporting Citizen Scientists with Automatic Species Identification Using Deep Learning Image Recognition Models." For the section on home-built and communal biolabs, I used De Lange et al., "Broadening Participation: 21st Century Opportunities for Amateurs in Biology Research," and Talbot, "The Rogue Experimenters." The information on and quotations from Sigrid Jakob come from her home page (sigridjakob.co); the piece she wrote is on the *Deep Funga Blog* of the Fungal Diversity Survey (fundis.org); and an interview with Josh McGinnis can be found on YouTube (youtu.be/i7C-2V1FPcI). She gave me permission to quote from her interview and home page, and she proofread the text on her lab.

CHAPTER 4: LABS FOR EVERYONE

The DIYbio listings can be consulted at sphere.diybio.org, and additional community ecology projects can be found at publiclab.org. The Barcoding the Harbor project is explained at bugssonline.org/group-projects/barcoding-the-harbor. I exchanged emails with the BUGSS director, Lisa

Scheifele, for the latest information on this project. The website of Genspace is genspace.org, and I took quotations from the video on that site. The interview with Jan-Maarten Luursema took place by phone on March 8, 2023; he also read and approved the text about his work. The website of Nature Lab is naturelab.risd.edu. The paragraph on Public Lab is based on what Andrew Quitmeyer told me and what is available on publiclab .org/about and the Wikipedia page on the 2010 BP oil spill. Other similar global initiatives are Hackteria and GOSH. The information on the Foldscope is from the February 17, 2023, episode of the podcast *Radiolab,* as well as from foldscope.com and foldscope.com/pages/microcosmos. Some smartphone applications and gadgets were taken from the item by Julien Bobroff at youtu.be/8ZZBCg317EA and from Cartwright, "Technology: Smartphone Science." The iSpex add-on is at www.ispex.nl/en and information on the SmidgION add-on is available at nanoporetech.com /products/smidgion. My interviews with Andrew Quitmeyer took place in person at the National University of Singapore on November 12, 2018, and by email on December 16, 2021, and March 15–21, 2023. Also, I used Quitmeyer, "Digital Naturalism: Interspecies Performative Tool Making for Embodied Science"; Quitmeyer et al., *Hacking the Wild;* and Quitmeyer et al., "Multidisciplinary Column: An Interview with Andrew Quitmeyer," as well as two other interviews with him, viz. Lang, "The Digital Naturalist," and Fong, "Andrew Quitmeyer's Wonderfully Weird World of Digital Naturalism." The ant traffic taster is outlined at https://www.instructables .com/Fiber-Optic-Jungle-Insect-Traffic-Taster. Andrew Quitmeyer read and approved the texts about his work. Jan-Maarten Luursema's 3D models are described at github.com under his profile, J4n-M44rt3n.

CHAPTER 5: VIRTUAL ACADEMIA

The visits of Crown Prince (later Emperor) Akihito to the Leiden museum and the imperial family's zoological hobbies are described in more detail in Schilthuizen and Vonk, *Wie wat bewaart* and based on Eisenstodt, "Behind the Chrysanthemum Curtain," Fransen and Van Oijen, "L. B. Holthuis, 'the Institutional Memory' of the Leiden Museum," Vervoort et al., "Personal Recollections of Lipke Bijdeley Holthuis," Alsemgeest and Fransen, *In krabbengang door kreeftenboeken* (pp. 183–184), and Datema,

"Japanse keizer krijgt geen warm welkom." A recent publication by him is Akihito and Ikeda, "Descriptions of Two New Species of *Callogobius* (Gobiidae) Found in Japan." The estimate of 30,000 scientific journals comes from publishingstate.com/how-many-academic-journals-are-there-in-the-world/2021. An analysis of the growth and change of character in academic publishing is Tenopir and King, "The Growth of Journals Publishing." A wonderful episode on Sci-Hub and its founder is available on a *Radiolab* podcast (April 7, 2023). Our MOOC is at https://www.coursera.org/learn/evolution-today and the Transmitting Science platform is at transmittingscience.com.

CHAPTER 6: ON CHESIL BEACHCOMBING

Details on the pebbles incident are taken from Kennedy, "McEwan Pebbles Back on Beach." My interview with E. O. Wilson is published as Schilthuizen, "In de ban van het kleine." The English book titles I mention are transliterations of the Dutch books *Dwalend ginds en her, Het jongens radioboek, Zien is kennen,* and *Welke kever is dat?* Daan Vestergaard was a biology teacher in the Stedelijk Gymnasium, Schiedam. The quotation is from Van der Wiel, *Welke kever is dat?* The numbers of species of *Megasternum* and *Cercyon* are taken from Fauna Europaea (www.fauna-eu.org). The anecdote about Schlegel and Brehm is from Schlegel, "Levensschets van Hermann Schlegel." The paragraph about the snap decisions that a collector needs to make is a paraphrased translation from Schilthuizen and Vonk, *Wie wat bewaart.* The details on the microhylid frogs are taken from amphibiaweb.org. Our work on using collection specimens to study snail shell evolution and bird cherry herbivory are in Ożgo and Schilthuizen, "Evolutionary Change in *Cepaea nemoralis* Shell Colour over 43 Years," Ożgo et al., "Inferring Microevolution from Museum Collections and Resampling: Lessons Learned from *Cepaea*," and Schilthuizen et al., "Incorporation of an Invasive Plant into a Native Insect Herbivore Food Web." Details on private collectors having contributed to the holdings of Naturalis Biodiversity Center are in Schilthuizen and Vonk, *Wie wat bewaart.* The Colorado study is from Gezon et al., "The Effect of Repeated, Lethal Sampling on Wild Bee Abundance and Diversity." The quantity of insects "consumed" by entomologists and bats has been estimated as five thousand per entomologist per year, and six thousand to eight thousand per bat per day (Griffin et al., "The Echolocation of Flying Insects by Bats"). A study of insects from light fixtures is Elbrecht et al., "A Bright Idea: Metabarcoding Arthropods from Light Fixtures."

CHAPTER 7: DEAD BUG BECOMES SPECIMEN

A good introduction to insect pinning is provided by *Science Friday*: www.sciencefriday.com/articles/insect-pinning-class. The alarms raised over the decline in donated collections are at https://whyy.org/segments/museums-arent-getting-as-many-animal-specimens-scientists-say-thats-bad and Fischer et al., "Decline of Amateur Lepidoptera Collectors Threatens the Future of Specimen-Based Research." Parts of the Packham-Attenborough interview I took from the *Yorkshire Post,* at https://www.yorkshirepost.co.uk/news/environment/young-naturalists-extinct-britain-1878883.

CHAPTER 8: HIDDEN RICHES

I took the information about Dyar's life mainly from Kelly, "Inside the Tunnels of Washington's Mole Man, Harrison G. Dyar," and Smith, "The Bizarre Tale of Tunnels, Trysts and Taxa of a Smithsonian Entomologist," but the best source is, of course, the biography by Epstein, *Moths, Myths, and Mosquitoes*. Some of Dyar's specimens can be found at https://bionomia.net/Q3127750. An account of recent hobby tunnelers, including the TikTok sensation Tunnel Girl, can be found in Xie, "Meet the DIY Diggers Who Can't Stop Making 'Hobby Tunnels.'" The children's book that meant so much to me as a child was Bomans and Poortvliet, *Pim, Frits en Ida, deel 8*. Recently my PhD student Auke-Florian Hiemstra tracked down a secondhand copy and gifted it to me. Some of my cave biology work is described in Schilthuizen et al., "Small-Scale Genetic Structuring in a Tropical Cave Snail and Admixture with Its Above-ground Sister Species." The original version of the trogloxene-troglophile-troglobite classification is in Schiner, "Fauna der Adelsberger-, Luegger-, und Magdalenen Grotte." The term "simple natural laboratories" is from Poulson and White, "The Cave Environment." The Pilgrim fathers in Leiden joined the *Mayflower,* which sailed from Plymouth, England, in 1620, and from there settled in

North America. The urban caving trip with Jasper van Kouwen took place on May 24, 2023. More about the underground spaces in Leiden can be found on his website, oudheidsfabriek.nl. The underground spaces of Catania are described in Bonaccorso et al., "Cavities and Hypogeal Structures of the Historical Part of the City of Catania." My information on the Odesa and Paris catacombs is mostly from McFarlane, "The Invisible City beneath Paris," Kovtun, "Подземные сооружения Одессы и Одесской области: сборник материалов 1-й научно-практической конференции (Одесса 11–12 ноября 2017 г.).—Одесса," Kovtun et al., "Invasive Landsnail *Oxychilus translucidus* (Stylommatophora, Zonitidae) in the Catacombs of Odesa (Ukraine)," Deli et al., "Предварительные данные о видовом составе пауков (Araneae) катакомб г. Одесса (Украина)," Drebet, "Monitoring of Bats in Key Wintering Shelters of the Northern Black Sea Region (Ukraine)," Sidorov and Kovtun, "*Synurella odessana* sp. n. (Crustacea, Amphipoda, Crangonyctidae): First Report of a Subterranean Amphipod from the Catacombs of Odessa and Its Zoogeographic Importance," and the entry at showcaves .com. Further details were supplied by email by Robert Vargovitsh and Oleg Kovtun. Lebreton and Héas, "La spéléologie urbaine: Une communauté secrète de cataphiles," gives a sociological account of the Parisian cataphiles. The Mumbai underground eel is described in Praveenraj et al., "*Rakthamichthys mumba*, a New Species of Hypogean Eel (Teleostei: Synbranchidae) from Mumbai, Maharashtra, India." The correspondence with Praveenraj Jayasimhan took place on April 4, 2023; he also checked and corrected the text I wrote about *Rakthamichthys mumba*. The information on *Niphargus aquilex* is from Aaf Verkade (April 2, 2023, by email), Pieters, "Een blinde vlokreeft maakte als 'grottenbeest' furore in een put," and Holthuis, "Notities betreffende Limburgse Crustacea. I. *Atyaephyra desmarestii* (Millet). II. *Niphargus aquilex* Schioedte." Most of the information on urban deep infrastructure is from Greene and Breisch, "Caverns of Concrete: Urban Karst Is a Challenging Frontier of Urban Entomology," and the report of *Aedes aegypti* in Washington, D.C., is from Lima et al., "Evidence for an Overwintering Population of *Aedes aegypti* in Capitol Hill Neighborhood, Washington, DC." The section on microbiomes on urban surfaces was based on Fomina et al., "Rock-Building Fungi," Hervé et al., "Aquatic Urban Ecology at the Scale of a Capital: Community Structure and Interactions in Street Gutters," Hervé and Lopez, "Analysis of

Interdomain Taxonomic Patterns in Urban Street Mats," Johnsen et al., "Quantification of Small-Scale Variation in the Size and Composition of Phenanthrene-Degrader Populations and PAH Contaminants in Traffic-Impacted Topsoil," Viles and Gorbushina, "Soiling and Microbial Colonisation on Urban Roadside Limestone: A Three-Year Study in Oxford, England," Gaylarde et al., "Epilithic and Endolithic Microorganisms and Deterioration on Stone Church Facades Subject to Urban Pollution in a Sub-tropical Climate," and Prieto-Barajas et al., "Microbial Mat Ecosystems: Structure Types, Functional Diversity, and Biotechnological Application." I took further details on the structure of microbial mats from Al-Thani et al., "Community Structure and Activity of a Highly Dynamic and Nutrient-Limited Hypersaline Microbial Mat in Um Alhool Sabkha, Qatar," and from the Wikipedia page on microbial mats.

CHAPTER 9: NOV. SPEC.

The second paragraph of this chapter is an edited English translation of a section in Schilthuizen and Vonk, *Wie wat bewaart*. The new and relict species from urban locations are in Gobster, "Alternative Approaches to Urban Natural Areas Restoration: Integrating Social and Ecological Goals," Wang et al., "Contribution to the Taxonomy of the Genus *Lycodon* H. Boie in Fitzinger, 1827 (Reptilia: Squamata: Colubridae) in China, with Description of Two New Species and Resurrection and Elevation of *Dinodon septentrionale chapaense* Angel, Bourret, 1933," Longino and Booher, "Expect the Unexpected: A New Ant from a Backyard in Utah," Baur et al., "Morphometric Analysis and Taxonomic Revision of *Anisopteromalus* Ruschka (Hymenoptera: Chalcidoidea: Pteromalidae): An Integrative Approach," and Van Achterberg et al., "A New Parasitoid Wasp, *Aphaereta vondelparkensis* sp. n.(Braconidae, Alysiinae), from a City Park in the Centre of Amsterdam."

CHAPTER 10: URBAN ISLANDS

Our paper on island biogeography in snails from offshore islands of northern Borneo is Schilthuizen et al., "Species Diversity Patterns in Insular Land Snail Communities of Borneo." The paragraphs on Surtsey are based on what I wrote about this island in my book *The Loom of Life*. The expeditions in greenspaces of Amsterdam were done by Taxon Expeditions between 2019 and 2024; the reports can be found at openresearch .amsterdam.nl. The urban island biogeography papers on Kunming and Bracknell are Gao et al., "Drivers of Spontaneous Plant Richness Patterns in Urban Green Space within a Biodiversity Hotspot," and Helden and Leather, "Biodiversity on Urban Roundabouts—Hemiptera, Management and the Species-Area Relationship"; the one on vacant lots is Crowe, "Lots of Weeds: Insular Phytogeography of Vacant Urban Lots." The quotations from Simon Leather are from iale.uk/roundabouts-can-be-so-much -more-just-traffic-calming-devices. More about the project Expedition Backyard can be found at https://taxonfoundation.com/project/expedition -backyard/?lang=en. A background on the SLoSS debate is Tjørve, "How to Resolve the SLOSS Debate: Lessons from Species-Diversity Models." The example of green corridors in Beijing is from Li et al., "Comprehensive Concept Planning of Urban Greening Based on Ecological Principles: A Case Study in Beijing, China." The scientific papers on the conservation value (or not) of fragmentation per se are Fahrig, "Ecological Responses to Habitat Fragmentation Per Se," Fahrig et al., "Is Habitat Fragmentation Bad for Biodiversity?," and Fletcher et al., "Is Habitat Fragmentation Good for Biodiversity?"

CHAPTER 11: INVOLUNTARY SLAUGHTER

The opening scene of this chapter was adapted from a column I wrote for the Dutch biologists' newsletter *Bionieuws* (July 16, 2019). The text on the Roadside Memorial Project was taken from lawatsonart.com. Desmond's text is from Desmond, "Requiem for Roadkill: Death and Denial on America's Roads," and the work of Brian Collier can be found at briancollier .net. Bram Koese's roadkill observations can be retrieved from waarneming .nl (the Dutch branch of Observation International) by searching for the location "Aarlanderveen—Polder Westzijde Aarlanderveen—Ziendeweg (Zuid-Holland)." Details on the 642 roadside shrines were taken from the media coverage of May 2021 in the newspapers *Algemeen Dagblad* and *De Nieuwkoper*, the radio stations NPO Radio 1 and Omroep West, and an online interview I did with Bram Koese between November 21 and 24,

2021. Bram Koese also proofread the text about his project. Dr. Splatt's school project can be found at roadkill.edutel.com and http://rjhroadkill.weebly.com. I also used an item from the June 27, 2018, issue of the *Eagle-Tribune*. The Belgian, Spanish, and Israeli citizen science roadkill mapping projects can be found, respectively, at www.dierenonderdewielen.be, observation.org/projects/38, and www.jpost.com/Business-and-Innovation/Tech/SPNI-Waze-identify-most-dangerous-roads-for-animals-497206. The scientific sources with continental and global estimates of roadkill statistics are Grilo et al., "Roadkill Risk and Population Vulnerability in European Birds and Mammals" and "Conservation Threats from Roadkill in the Global Road Network," Jensen et al., *Bright Green Lies: How the Environmental Movement Lost Its Way and What We Can Do about It*, Loss et al., "Estimation of Bird-Vehicle Collision Mortality on U.S. roads," and Seiler and Helldin, "Mortality in Wildlife due to Transportation." The carcass retention paper is Santos et al., "How Long Do the Dead Survive on the Road? Carcass Persistence Probability and Implications for Road-Kill Monitoring Surveys." Papers on insect roadkill are Baxter-Gilbert et al., "Road Mortality Potentially Responsible for Billions of Pollinating Insect Deaths Annually," Messenger, "Trillions of Insects Killed by Cars Every Year, Says Study," and Muñoz et al., "Effects of Roads on Insects: A Review." The figures on the extinction risk of specific species of mammals and bird come from Grilo et al., "Roadkill Risk and Population Vulnerability in European Birds and Mammals" and "Conservation Threats from Roadkill in the Global Road Network." The one on the intentional killing of reptiles is Ashley et al., "Incidence of Intentional Reptile-Vehicle Collisions"; Ashley also used a second control, a plastic cup, which for clarity I have not mentioned in the text. The early records of roadkill from 1897, 1920, 1924, and the 1950s are, respectively, from Knutson, *Flattened Fauna*, Grinnell, May 5 notes, Stoner, "The Toll of the Automobile," and Mörzer Bruijns, "Faunasterfte door verkeer." Information on OpenStreetMap was taken from openstreetmap.org and from wiki.openstreetmap.org; the paper using the data to calculate roadless areas is Ibisch et al., "A Global Map of Roadless Areas and Their Conservation Status." The estimations of road length in the next decades are from Alamgir et al., "Economic, Socio-Political and Environmental Risks of Road Development in the Tropics," and Lawton, "Road Kill." The barn owl "hostile design" can be found at: tinyurl.com/488ubfm3. Information on polyethylene fencing can be found at kingrootbarrier.com. An English-language page about the hundred-meter-wide ecoduct over the A4 in the Netherlands is at https://www.wikipe.wiki/wiki/nl/Eco-aquaduct_Zweth_en_Slinksloot. The "Band-Aid" quotation is from Laurance, "If You Can't Build Well, Then Build Nothing at All," the Butterfly Effect product is described at https://solarix-solar.com/the-butterfly-effect-an-ultra-light-web-over-the-highway/?lang=en, and the community anti-road building project in the Netherlands can be found at https://tinyurl.com/6zevzhhn.

CHAPTER 12: IT'S A TRAP!

Hsu's photos are at www.tinyurl.com/ct3p5g. His words describing the photos are based on their Dutch translation by Moeliker, *De eendenman*. The social media post is by @bitesizevegan on Instagram of April 3, 2016, and the ScienceBlogs post is at https://scienceblogs.com/grrlscientist/2008/10/31/one-of-lifes-tiny-dramas. The idea that bird flu spreads via necrophilia is from a podcast interview with Mardik Leopold (episode 41 in *Toekomst voor Natuur*, https://www.vlinderstichting.nl/toekomstvoornatuur). The full text of the limerick on which the term "Davian" is based (or at least one version of it) is: "There once was a hermit named Dave / Who kept a dead whore in his cave. / 'I know it's a sin,' / He said with a grin, / "But think of the money I save!" Examples of Davian behavior in birds, mammals, and amphibians are in Dale, "Necrophilic Behaviour, Corpses as Nuclei of Resting Flock Formation, and Road-Kills of Sand Martins *Riparia riparia*," Dickerman, "'Davian Behavior Complex' in Ground Squirrels," Lehner, "Avian Davian Behavior," Meshaka, "Anuran Davian Behavior: A Darwinian Dilemma," and Moeliker, *De eendenman*. The house sparrow example is from Simmons, "Bizarre Behaviour and Death of Male House Sparrow." The study on slow worms on bicycle tracks is in Bijlsma, *Kerken van goud, dominees van hout*. Roadkill caused by attraction to roads because of carrion, salt, or flat migration routes is mentioned in, for example, Hill et al., "A Review of Ecological Factors Promoting Road Use by Mammals," and Lehtonen et al., "High Road Mortality during Female-Biased Larval Dispersal in an Iconic Beetle." The studies on animals trapped in bottles and cans are, for mammals, Debernardi et al., "Small Mammals Found in Discarded Bottles in Alpine and Pre-Alpine Areas of NW-Italy," and Arrizabalaga et al., "Small Mammals in Discarded Bottles: A New World Record," and, for beetles, Romiti et al., "Quantifying the Entrapment Effect of Anthropogenic Beach Litter on Sand-Dwelling Beetles According to the EU Marine Strategy Framework Directive." The community science projects on bottles as unintended animal traps are Moates, "Small Mammal Mortality in Discarded Bottles and Drinks Cans—A Norfolk-Based Field Study in a Global Context," and Kolenda et al., "Online Media Reveals a Global Problem of Discarded Containers as Deadly Traps for Animals." The details on Moates's project were further obtained from the March 26 and 27, 2018, issues of the *Eastern Daily Press* (www.edp24.co.uk), the *Daily Mail* (www.dailymail.co.uk), and the *Express and Star* (www.expressandstar.com). A small biography of Bernhard van Vondel is in Plaisier, "De collectiebeheerders aflevering 8: Bernhard van Vondel." The work on polarizing surfaces as evolutionary traps is from the following: Horváth et al., "Ecological Traps for Dragonflies in a Cemetery: The Attraction of *Sympetrum* Species (Odonata: Libellulidae) by Horizontally Polarizing Black Gravestones" (gravestones); Horváth et al., "Reducing the Maladaptive Attractiveness of Solar Panels to Polarotactic Insects" (solar panels); Horváth and Zeil, "Kuwait Oil Lakes as Insect Traps" (oil lakes); Van Vondel, "Another Case of Water Beetles Landing on a Red Car Roof," Schilthuizen, "Carpoelen," and Kriska et al., "Why Do Red and Dark-Coloured Cars Lure Aquatic Insects? The Attraction of Water Insects to Car Paintwork Explained by Reflection–Polarization Signals" (car roofs); Malik et al., "Imaging Polarimetry of Glass Buildings: Why Do Vertical Glass Surfaces Attract Polarotactic Insects?" (glass windows); and Kriska et al., "Why Do Mayflies Lay Their Eggs En Masse on Dry Asphalt Roads? Water-Imitating Polarized Light Reflected from Asphalt Attracts Ephemeroptera" (asphalt). The studies that show various mitigating designs of solar panels are Horváth et al., "Reducing the Maladaptive Attractiveness of Solar Panels to Polarotactic Insects," Black and Robertson, "How to Disguise Evolutionary Traps Created by solar panels," and Fritz et al., "Bioreplicated Coatings for Photovoltaic Solar Panels Nearly Eliminate Light Pollution That Harms Polarotactic Insects." The other examples of evolutionary traps in the final section are from Semel and Sherman, "Intraspecific Parasitism and Nest-Site Competition in Wood Ducks," and Schlaepfer et al., "Ecological and Evolutionary Traps." My previous book about urban evolution is Schilthuizen, *Darwin Comes to Town*. The information on birds being attracted to solar panels is from email correspondence with Bruce Robertson, December 9–11, 2021.

CHAPTER 13: ANIMAL ARCHITECTS OF THE ANTHROPOCENE

The Grachtwacht can be found at www.degrachtwacht.nl. Their papers I mention are Hiemstra et al., "The Effects of COVID-19 Litter on Animal Life" and "Birds Using Artificial Plants as Nesting Material," and Tasseron et al., "Plastic Hotspot Mapping in Urban Water Systems." The interview with Hiemstra is a combination of various meetings and conversations and is also based on his lecture of March 24, 2023, in the Leiden University course Urban Ecology and Evolution. Nozeman and Sepp's book, *Nederlandsche vogelen*, can be found online at https://www.kb.nl/ontdekken-bewonderen/topstukken/nederlandsche-vogelen. Auke-Florian Hiemstra read and approved my text on his work. The paper on human-made mass versus biomass is by Elhacham et al., "Global Human-Made Mass Exceeds All Living Biomass," and the estimate of annual plastic production is from Williams and Rangel-Buitrago, "The Past, Present, and Future of Plastic Pollution." The crows that make nests from clothes hangers are decribed in, for example, Lim and Lim, "Hachioji's Environment." Hermit crabs using plastic "shells" are shown at, for example, https://okinawanaturephotography.com/tag/offering-hermit-crabs-shells-instead-of-plastic and discussed in Jagiello et al., "The Plastic Homes of Hermit Crabs in the Anthropocene." The paper on birds using antibird spikes as nest materials is Hiemstra et al., "Bird Nests Made from Anti-Bird Spikes." The work on cigarette butts in bird nests is Suárez-Rodríguez et al., "Incorporation of Cigarette Butts into Rests Reduces Nest Ectoparasite Load in Urban Birds: New Ingredients for an Old Recipe?," and the paper on plastic nest decoration in black kites is Sergio et al., "Raptor Nest Decorations Are a Reliable Threat against Conspecifics." Quotations from Diana Lelonek are from her interview published on the website http://newalphabetschool.hkw.de.

CHAPTER 14: THE ACCIDENTAL ECOSYSTEM

The visit to the Belgrade Museum of Contemporary Art took place on April 21, 2023. My information on the invasive stink bug comes from Hamilton et al., "*Halyomorpha halys* (Stål)." The discourse around three Asian insects in Belgrade is largely taken from Filipović, "Three Bugs in the City: Urban Ecology and Multispecies Relationality in Postsocialist Belgrade." I interviewed Jovana Bila Dubaić on April 20, 2023, and she gave me further details on Belgrade development projects via Instagram messenger on January 12 and 13, 2024. Details on the Linear Park are from https://nedavimobeograd.rs/linijski-park-ili-linijska-prevara. The example from the San Francisco Bay is from Cohen and Carlton, "Accelerated Invasion Rate in a Highly Invaded Estuary," and the one from Chile is from Santilli et al., "Exotic Species Predominates in the Urban Woody Flora of Central Chile." The example of New Zealand birds is from Van Heezik et al., "Diversity of Native and Exotic Birds across an Urban Gradient in a New Zealand City" (I took the greater than 50 percent figure from their table 2). The quotations from Ingo Kowarik are from "Novel Urban Ecosystems, Biodiversity, and Conservation." The Mexico City hummingbird study is described in Marín-Gómez et al., "Assessing Ecological Interactions in Urban Areas Using Citizen Science Data: Insights from Hummingbird–Plant Meta-Networks in a Tropical Megacity." The horse chestnut study is from Pocock and Evans, "The Success of the Horse-Chestnut Leaf-Miner, *Cameraria ohridella*, in the UK Revealed with Hypothesis-Led Citizen Science." The sources on waterfowl eating invasive crayfish in the Netherlands are all online Dutch sources, viz., https://www.naturetoday.com/intl/nl/nature-reports/message/?msg=29580 and https://sleutelstad.nl/2023/07/27/leidse-mantelmeeuwen-maken-stoofpotje-van-kreeft. The study on beetles feeding on American bird cherry is Schilthuizen et al., "Incorporation of an Invasive Plant into a Native Insect Herbivore Food Web."

CHAPTER 15: ON THE ORIGIN OF URBAN SPECIES

Most of the examples in this chapter are based on research that has appeared since Schilthuizen, *Darwin Comes to Town,* came out. The stories about urban *Anolis* lizards are based on Winchell et al., "Linking Locomotor Performance to Morphological Shifts in Urban Lizards," Falvey et al., "The Finer Points of Urban Adaptation: Intraspecific Variation in Lizard Claw Morphology," Howell et al., "Geometric Morphometrics Reveal Shape Differences in the Toes of Urban Lizards," and on several blog pieces by Emmanuel D'Agostino, Bailey Howell, and Cleo Falvey posted on urbanevolution-litc.com. For background on *Anolis*, I also used www.anoleannals.org and Losos and Schneider, "*Anolis* Lizards." Kristin Winchell read and checked my text. The paper on dandelion seeds is Arathi, "A Comparison of Dispersal Traits in Dandelions Growing in Urban landscape and open Meadows." Most of the results of the Dutch research projects on dandelion seeds (kindly communicated to me by Barbara Gravendeel, Niels Kerstes, and Kitty Vijverberg in emails of October 2, 4, and 5, 2022, respectively) are still unpublished, but some information can be found at https://www.naturetoday.com/intl/en/nature-reports/message/?msg=28132 and https://www.bnnvara.nl/vroegevogels/artikelen/doe-mee-met-paardenbloemenonderzoek. The work on the Australian water dragon is in Littleford-Colquhoun et al., "Archipelagos of the Anthropocene: Rapid and Extensive Differentiation of Native Terrestrial Vertebrates in a Single Metropolis." Some general information on thermal evolution in *Cepaea* snails, in particular in urban areas, is in Ożgo and Schilthuizen, "Evolutionary Change in *Cepaea nemoralis* Shell Colour over 43 Years," Schilthuizen, "Rapid, Habitat-Related Evolution of Land Snail Colour Morphs on Reclaimed Land," and Kerstes et al., "Snail Shell Colour Evolution in Urban Heat Islands Detected via Citizen Science." More about the acorn ant research and background can be found at https://www.diamond-lab.org and in Diamond et al., "Rapid Evolution of Ant Thermal Tolerance across an Urban-Rural Temperature Cline," Chick et al., "Urban Heat Islands Advance the Timing of Reproduction in a Social Insect," and Martin et al., "Evolution, Not Transgenerational Plasticity, Explains the Adaptive Divergence of Acorn Ant Thermal Tolerance across an Urban–Rural Temperature Cline" and "In a Nutshell, a Reciprocal Transplant Experiment Reveals Local Adaptation and Fitness Trade-offs in Response to Urban Evolution in an Acorn-Dwelling Ant." The "drunken staggering" quotation is from Yin, "What Makes a City Ant? Maybe Just 100 Years of Evolution." In the section about ALAN, the information on the spiders of Turin's archways comes from Mammola et al., "Artificial Lighting Triggers the Presence of Urban Spiders and Their Webs on Historical Buildings," and https://www.centrorestaurovenaria.it. The experiments on the evolution of light attraction in *Steatoda* are described in Czaczkes et al., "Reduced Light Avoidance in Spiders from Populations in Light-Polluted Urban Environments," and the study on the Viennese bridge spiders is Heiling, "Why Do Nocturnal Orb-Web Spiders (Araneidae) Search for Light?" The quotation from Florian Altermatt comes from email correspondence with him on December 9 and 10, 2022, and his research is described in Altermatt and Ebert, "Reduced Flight-to-Light Behaviour of Moth Populations Exposed to Long-Term Urban Light Pollution."

CHAPTER 16: WE ARE A NODE

The ecosystem in and around my house in Kota Kinabalu was described in much more detail in my book *The Loom of Life*. A paper on the problem-solving abilities of raccoons is Daniels et al., "Behavioral Flexibility of a Generalist Carnivore". The papers on cockatoos opening bins are Klump et al., "Innovation and Geographic Spread of a Complex Foraging Culture in an Urban Parrot" and "Is Bin-Opening in Cockatoos Leading to an Innovation Arms Race with Humans?" The paper on boldness in Australian water dragons is Baxter-Gilbert et al., "Bold New World: Urbanization Promotes an Innate Behavioral Trait in a Lizard." Information on the distribution of *Aedes aegypti* is from Kraemer et al., "The Global Distribution of the Arbovirus Vectors *Aedes aegypti* and *Ae. albopictus*," and information on the evolution of human biting is from Rose et al., "Climate and Urbanization Drive Mosquito Preference for Humans." The insights into *Flavivirus* evolution are from Gould et al., "Origins, Evolution, and Vector/Host Coadaptations within the Genus *Flavivirus*," while an overview of potential human evolution as a result of dengue infection can be found in Sierra et al., "*OSBPL10*, *RXRA* and Lipid Metabolism Confer African-Ancestry Protection against Dengue Haemorrhagic Fever in Admixed Cubans," and Oliveira et al., "Host Ancestry and Dengue Fever: From Mapping of Candidate Genes to Prediction of Worldwide Genetic Risk." The analysis of evolution in immune genes is from Daub et al., "Evidence for Polygenic Adaptation to Pathogens in the Human Genome," and the comparison of ancient and modern European genomes is from Chekalin et al., "Changes in Biological Pathways during 6,000 Years of Civilization in Europe." General ideas about the evolvability of modern humans are based on Hawks et al., "Recent Acceleration of Human Adaptive Evolution," and Milot and Stearns, "Selection on Humans in Cities." The price of a human genome is from the National Institutes of Health's website genome.gov, and the Biobank results are from Gibbons, "Spotting Evolution among Us." The example of sawflies on Raywood ash is based on observations told to me by Fons Verheyde.

CHAPTER 17: SPEAK SOFTLY AND CARRY A BIG STICK INSECT

In the account of my run-in with the police for destroying the property of my local soccer club, the name of my friend Isaac has been changed. My visit to Istanbul took place from January 16 to 26, 2022. Information on the greenspace area of cities comes from http://www.worldcitiescultureforum.com/data/of-public-green-space-parks-and-gardens, from my interview with Emrah Çoraman on January 19, 2022, and from communications with Emrah Çoraman on May 16, 2022, that provided additional details on active and passive greenspaces in Istanbul. The timeline of the Gezi Parks protests is from the Wikipedia page en.wikipedia.org/wiki/Gezi_Park_protests, consulted throughout May 2022, and from my interview with Rana Söylemez on January 23, 2022. The information on the Istanbul *bostan* comes from my interview with Rana Söylemez, from Hattam, "In Istanbul's Ancient Gardens, a Battle for Future Harvests," and from the websites www.bostanhikayeleri.com and yedikulebostanlari.tumblr.com. The background on Validebağ Grove is from the websites www.validebag.org and www.birdingsites.eu, while the information on the recent threats and the volunteers' response is from Smith, "Protesters Sit Tight as President Erdogan's Trucks Roll In to Pave Over Validebag Grove in Istanbul," and Pişkin, "'This Is a Murder': Construction Equipment Enters İstanbul's Validebağ Grove." My interview with Cihan Babuccu at Validebağ took place on January 21, 2022. The current number of biodiversity hotspots is in Habel et al., "Final Countdown for

Biodiversity Hotspots." The text about Istanbul was checked by Cihan Babuccu, Emrah Çoraman, and Rana Söylemez.

CHAPTER 18: *TAK KENAL MAKA TAK CINTA*

The text about Bukit Kiara and its fireflies was based on discussions with Wan Yusoh during her visit to Leiden in December 2019, a videocall with Thary Goh on August 19, 2021, video lectures available on facebook.com /BukitKiaraFriends, the YouTube video at https://youtu.be/Z68s-OGHpos, Chan, "The World's Largest Firefly Can Be Found in Kuala Lumpur," and Mbugua et al., "Effects of Artificial Light on the Larvae of the Firefly *Lamprigera sp.* in an Urban City Park, Peninsular Malaysia." The two books I mention by Michael Rosenzweig are *Species Diversity in Space and Time* and *Win-Win Ecology: How the Earth's Species Can Survive in the Midst of Human Enterprise.*

CHAPTER 19: LET IT GROW

The details on the vegetation of Batu Caves come from, besides personal experience, Kiew and Abdul Rahman, "Plant Diversity Assessment of Karst Limestone, a Case Study of Malaysia's Batu Caves." Much on the ecology of green walls was taken from Francis, "Wall Ecology: A Frontier for Urban Biodiversity and Ecological Engineering," and Segal, *Ecological Notes on Wall Vegetation.* I met and spoke with Rob Francis during the "Urban Ecologies" workshop at the University of Cambridge on November 22, 2018; other details were provided by him via Twitter direct messages on June 24, 2021. More on ecosystem assembly on real and artificial islands can be found in the chapter "In Splendid Isolation" of my book *The Loom of Life.* Details on artists working on spontaneous vegetation in urban contexts, including Weinberger and Clément, were taken from Costa, "Ways of Engaging," and Oskam and Mota, "Design in the Anthropocene: Intentions for the Unintentional." I credit my friend Frank van Rooij for the idea to pre-create a pattern or an image in a spontaneously vegetated green wall. I obtained details on Douglas Boyes from his blog, douglasboyes.co.uk, the *Guardian*'s "Douglas Boyes Obituary," Clements,

"'My Son Took His Own Life—It's a Tragedy He Couldn't See How Much He Was Valued,'" and from his Twitter account, @diarsia, which his mother, Clare, still maintains. His scientific paper that I mention is Boyes et al., "Street Lighting Has Detrimental Impacts on Local Insect Populations." The work by Jonas Frei is *Stadtwildpflanzen*; I also obtained information from his website, foifacht.ch, and his Twitter account, @jdfrei. The discussion of Ton Denters's work draws mainly on his Twitter account (@urbane_natuur) and on Denters, *Stadsflora van de lage landen*; the details on artificial grass come from the September 11, 2022, column by Kees Moeliker on the Dutch radio program *Vroege Vogels*, the Wikipedia entries "AstroTurf" and "Artificial Turf," and Wolters et al., "Astroturf Seed Traps for Studying Hydrochory." The report on greens workers in the UK mowing astroturf lawns is from a BBC report on April 8, 2022 (https://www.bbc.com/news/uk-england-somerset-61043514). The rate of growth of the astroturf market came from http://alliedmarketresearch.com. The website for Hans van Lunteren's project is https://www.utrechtnatuurlijk.nl/toevalliggroen. An example of a botanic garden displaying pavement weeds is Heimanshof in Hoofddorp, the Netherlands. The website of the former company Gewildgroei is at www.gewildgroei.nl; I also got information from Vincent Wittenberg and from "Bennie Meek: Living Pavement." The quoted words of Diana Lelonek are from her interview on http://newalphabetschool.hkw.de.

CLOSER

The email correspondence with Neha Dsouza took place between July 12, 2021, and February 28, 2022; she gave me explicit permission to quote from her messages. The writer Yvonne Adhiambo Owuor first told me about the mosquito that gave comfort to political prisoner Patrick Onyango. Dr. Onyango himself then recounted the story to me via an exchange of direct messages on Twitter between June 14 and 19, 2023. The story of his and other activists' detainment and torture is told in Citizens for Justice, *We Lived to Tell*. I would like to add one exception to my pronouncement that we should not fight exotic species in cities: free-roaming domestic cats. Supplemented with food by humans, these predators reach densities that are unsustainably high. Under natural conditions, where the house cat's ancestor the European wildcat lives, you would find one cat per ten square kilometers. That is the home range a cat needs to catch enough prey. But house cats are fed by humans *and* permitted to leave the house and prowl the neighborhood to feed on wild birds, insects, small mammals, amphibians, and reptiles at densities several orders of magnitude higher, leading to a depletion of any wild urban animals large enough to fit into a kitty's fanged mouth. (The home range size of a wildcat is my estimate based on Anile et al., "Home-Range Size of the European Wildcat [*Felis silvestris silvestris*]: A Report from Two Areas in Central Italy.")

BIBLIOGRAPHY

Akihito and Y. Ikeda. "Descriptions of Two New Species of *Callogobius* (Gobiidae) Found in Japan." *Ichthyological Research* 69 (2022): 97–110.

Alamgir, M., M. J. Campbell, S. Sloan, M. Goosem, G. R. Clements, M. I. Mahmoud, and W. F. Laurance. "Economic, Socio-political and Environmental Risks of Road Development in the Tropics." *Current Biology* 27 (2017): R1130–R1140.

Alsemgeest, A., and C. Fransen. *In krabbengang door kreeftenboeken; de Bibliotheca Carcinologica L. B. Holthuis*. Leiden: Naturalis, 2016.

Altermatt, F., and D. Ebert. "Reduced Flight-to-Light Behaviour of Moth Populations Exposed to Long-Term Urban Light Pollution." *Biology Letters* 12, no. 4 (2016): art. 2016.0111.

Al-Thani, R., M. A. A. Al-Najjar, A. Munem Al-Raei, T. Ferdelman, N. M. Thang, I. Al Shaikh, M. Al-Ansi, and D. de Beer. "Community Structure and Activity of a Highly Dynamic and Nutrient-Limited Hypersaline Microbial Mat in Um Alhool Sabkha, Qatar." *PLoS One* 9 (2014): art. e92405.

Anile, S., L. Bizzarri, M. Lacrimini, A. Sforzi, B. Ragni, and S. Devillard. "Home-Range Size of the European Wildcat (*Felis silvestris silvestris*): A Report from Two Areas in Central Italy." *Mammalia* 82 (2017): 1–11.

Arathi, H. S. "A Comparison of Dispersal Traits in Dandelions Growing in Urban Landscape and Open Meadows." *Journal of Plant Studies* 1 (2012): 40–46.

Arrizabalaga, A., L. M. González, and I. Torre. "Small Mammals in Discarded Bottles: A New World Record." *Galemys* 28 (2016): 63–65.

Ashley, E. P., A. Kosloski, and S. A. Petrie. "Incidence of Intentional Reptile-Vehicle Collisions." *Human Dimensions of Wildlife* 12 (2007): 137–143.

Baur, H., Y. Kranz-Baltensperger, A. Cruaud, J. Y. Rasplus, A. V. Timokhov, and V. E. Gokhman. "Morphometric Analysis and Taxonomic Revision of *Anisopteromalus* Ruschka (Hymenoptera: Chalcidoidea: Pteromalidae): An Integrative Approach." *Systematic Entomology* 39 (2014): 691–709.

Baxter-Gilbert, J. H., J. L. Riley, C. J. H. Neufeld, J. D. Litzgus, and D. Lesbarrères. "Road Mortality Potentially Responsible for Billions of Pollinating Insect Deaths Annually." *Journal of Insect Conservation* 19 (2015): 1029–1035.

Baxter-Gilbert, J., J. L. Riley, and M. J. Whiting. "Bold New World: Urbanization Promotes an Innate Behavioral Trait in a Lizard." *Behavioral Ecology and Sociobiology* 73 (2019): 105.

"Bennie Meek: Living Pavement." *Domus*, November 19, 2012.

Bijlsma, R. *Kerken van goud, dominees van hout.* Atlas Contact, 2021.

Black, T. V., and B. A. Robertson. "How to Disguise Evolutionary Traps Created by Solar Panels." *Journal of Insect Conservation* 24 (2020): 241–247.

Bomans, G., and R. Poortvliet. *Pim, Frits en Ida, deel 8: Diep onder de aarde*. Groningen: Dijkstra, 1968.

Bonaccorso, R., S. Grasso, E. L. Giudice, and M. Maugeri. "Cavities and Hypogeal Structures of the Historical Part of the City of Catania." *WIT Transactions on State-of-the-Art in Science and Engineering* 8 (2005): 197–223.

Boyes, D. H., D. M. Evans, R. Fox, M. S. Parsons, and M. J. Pocock. "Street Lighting Has Detrimental Impacts on Local Insect Populations." *Science Advances* 7 (2021): art. eabi8322.

Brenner, K. *Nature Obscura: A City's Hidden Natural World*. Seattle: Mountaineers Books, 2020.

Cartwright, J. "Technology: Smartphone Science." *Nature* 531 (2016): 669–671.

Chan, E. "The World's Largest Firefly Can Be Found in Kuala Lumpur." *Star*, August 12, 2017.

Chekalin, E., A. Rubanovich, T. V. Tatarinova, A. Kasianov, N. Bender, M. Chekalina, K. Staub, N. Koepke, F. Rühli, S. Bruskin, and I. Morozova. "Changes in Biological Pathways during 6,000 Years of Civilization in Europe." *Molecular Biology and Evolution* 36 (2019): 127–140.

Chick, L. D., S. A. Strickler, A. Perez, R. A. Martin, and S. E. Diamond. "Urban Heat Islands Advance the Timing of Reproduction in a Social Insect." *Journal of Thermal Biology* 80 (2019): 119–125.

Citizens for Justice. *We Lived to Tell: The Nyayo House Story*. Bonn: Friedrich Ebert Stiftung, 2003.

Clements, L. "'My Son Took His Own Life—It's a Tragedy He Couldn't See How Much He Was Valued.'" *Wales Online*, August 28, 2022, https://www.walesonline.co.uk/news/my-son-took-life-its-24851389.

Cohen, A. N., and J. T. Carlton. "Accelerated Invasion Rate in a Highly Invaded Estuary." *Science* 279 (1998): 555–558.

Costa, A. M. D. G. "Ways of Engaging: Re-Assessing Effects of the Relationship between Landscape Architecture and Art in Community Involvement and Design Practice." PhD diss., Universidade de Lisboa, Portugal, 2018.

Crowe, T. M. Lots of Weeds: "Insular Phytogeography of Vacant Urban Lots." *Journal of Biogeography* 6 (1979): 169–181.

Czaczkes, T. J., A. M. Bastidas-Urrutia, P. Ghislandi, and C. Tuni. "Reduced Light Avoidance in Spiders from Populations in Light-Polluted Urban Environments." *Science of Nature* 105 (2018): 1–5.

Dale, S. "Necrophilic Behaviour, Corpses as Nuclei of Resting Flock Formation, and Road-Kills of Sand Martins *Riparia riparia*." *Ardea* 89 (2001): 545–547.

Daniels, S. E., R. E. Fanelli, A. Gilbert, and S. Benson-Amram. "Behavioral Flexibility of a Generalist Carnivore." *Animal Cognition* 22 (2019): 387–396.

Datema, D. "Japanse keizer krijgt geen warm welkom." *De Dag van Toen*, 2017, https://www.dagvantoen.nl/japanse-keizer-krijgt-geen-warm-welkom.

Daub, J. T., T. Hofer, E. Cutivet, I. Dupanloup, L. Quintana-Murci, M. Robinson-Rechavi, and L. Excoffier. "Evidence for Polygenic Adaptation to Pathogens in the Human Genome. *Molecular Biology and Evolution* 30 (2013): 1544–1558.

De Lange, O., C. Youngflesh, A. Ibarra, R. Perez, and M. Kaplan. "Broadening Participation: 21st Century Opportunities for Amateurs in Biology Research." *Integrative and Comparative Biology* 61 (2021): 2294–2305.

Debernardi, P., E. Patriarca, A. Perrone, M. Cantini, and B. Chiarenzi. "Small Mammals Found in Discarded Bottles in Alpine and Pre-alpine Areas of NW-Italy." *Hystrix* 9 (1997): 51–55.

Deli, O. F., O. A. Kovtun, and K. K. Pronin. Предварительные данные о видовом составе пауков (Araneae) катакомб г. Одесса (Украина) [Preliminary data on the species composition of spiders (Araneae) in the catacombs of Odessa (Ukraine)]. In Биоспелеологические исследования в России и сопредельных государствах [Biospeleological Research in Russia and Neighboring Countries], 25–29. Moscow: Russian Academy of Sciences, 2017.

Denters, T. *Stadsflora van de lage landen*. Amsterdam: Fontaine Uitgevers, 2020.

Desmond, J. "Requiem for Roadkill. Death and Denial on America's Roads." In *Environmental Anthropology: Future Directions*, ed. H. Kopnina and E. Shoreman-Ouimet, 46–58. Routledge, 2013.

Diamond, S. E., L. Chick, A. B. E. Perez, S. A. Strickler, and R. A. Martin. "Rapid Evolution of Ant Thermal Tolerance across an Urban-Rural Temperature Cline." *Biological Journal of the Linnean Society* 121 (2017): 248–257.

Dickerman, R. W. "'Davian Behavior Complex' in Ground Squirrels." *Journal of Mammalogy* 41 (1960): 403.

"Douglas Boyes Obituary." *Guardian*, November 30, 2021, https://www.theguardian.com/environment/2021/nov/30/douglas-boyes-obituary.

Drebet, M. "Monitoring of Bats in Key Wintering Shelters of the Northern Black Sea Region (Ukraine)." *Theriologia Ukrainica* 23 (2022): 11–19.

Eisenstodt, G. "Behind the Chrysanthemum Curtain." *Atlantic*, November 1998.

Elbrecht, V., A. Lindner, L. Manerus, and D. Steinke. "A Bright Idea: Metabarcoding Arthropods from Light Fixtures." *PeerJ* 9 (2021): art. e11841.

Elhacham, E., L. Ben-Uri, J. Grozovski, Y. M. Bar-On, and R. Milo. "Global Human-Made Mass Exceeds All Living Biomass." *Nature* 588 (2020): 442–444.

Epstein, M. E. *Moths, Myths, & Mosquitoes: The Eccentric Life of Harrison G. Dyar, Jr.* Oxford University Press, 2016.

Fahrig, L. "Ecological Responses to Habitat Fragmentation Per Se." *Annual Review of Ecology, Evolution, and Systematics* 48 (2017): 1–23.

Fahrig, L., V. Arroyo-Rodríguez, J. R. Bennett, V. Boucher-Lalonde, E. Cazetta, D. J. Currie, F. Eigenbrod, et al. "Is Habitat Fragmentation Bad for Biodiversity?" *Biological Conservation* 230 (2019): 179–186.

Falvey, C. H., K. J. Aviles-Rodriguez, T. J. Hagey, and K. M. Winchell. "The Finer Points of Urban Adaptation: Intraspecific Variation in Lizard Claw Morphology." *Biological Journal of the Linnean Society* 131 (2020): 304–318.

Filipović, A. "Three Bugs in the City: Urban Ecology and Multispecies Relationality in Postsocialist Belgrade." *Contemporary Social Science* 16 (2021): 29–42.

Fischer, E. E., N. S. Cobb, A. Y. Kawahara, J. M. Zaspel, and A. I. Cognato. "Decline of Amateur Lepidoptera Collectors Threatens the Future of Specimen-Based Research." *BioScience* 71 (2021): 396–404.

Fletcher, R. J., R. K. Didham, C. Banks-Leite, J. Barlow, R. M. Ewers, J. Rosindell, et al. "Is Habitat Fragmentation Good for Biodiversity?" *Biological Conservation* 226 (2018): 9–15.

Fomina, M., E. P. Burford, S. Hillier, M. Kierans, and G. M. Gadd. "Rock-Building Fungi." *Geomicrobiology Journal* 27 (2010): 624–629.

Fong, C. "Andrew Quitmeyer's Wonderfully Weird World of Digital Naturalism." *Makery*, June 30, 2019, makery.info.

Francis, R. A. "Wall Ecology: A Frontier for Urban Biodiversity and Ecological Engineering." *Progress in Physical Geography* 35 (2011): 43–63.

Fransen, C. H. J. M., and M. J. P. van Oijen. "L. B. Holthuis, 'the Institutional Memory' of the Leiden Museum—Obituary." *Contributions to Zoology* 77 (2008): 201–204.

Frei, J. D. *Stadtwildpflanzen: 52 Ausflüge in die urbane Pflanzenwelt. Mit Hintergrundwissen zur Stadtvegetation.* Zurich: AT Verlag, 2022.

Fritz, B., G. Horváth, R. Hünig, Á. Pereszlényi, Á. Egri, M. Guttmann, M. Schneider, U. Lemmer, G. Kriska, and G. Gommard. "Bioreplicated Coatings for Photovoltaic Solar Panels Nearly Eliminate Light Pollution That Harms Polarotactic Insects." *PLoS ONE* 15 (2020): art. e0243296.

Gao, Z., K. Song, Y. Pan, D. Malkinson, X. Zhang, B. Jia, T. Xia, et al. "Drivers of Spontaneous Plant Richness Patterns in Urban Green Space within a Biodiversity Hotspot." *Urban Forestry and Urban Greening* 61 (2021): art. 127098.

Garber, S. D. *The Urban Naturalist*. New York: John Wiley & Sons, 1987.

Gaylarde, C., J. A. Baptista-Neto, A. Ogawa, M. Kowalski, S. Celikkol-Aydin, and I. Beech. "Epilithic and Endolithic Microorganisms and Deterioration on Stone Church Facades Subject to Urban Pollution in a Sub-tropical Climate." *Biofouling* 33 (2017): 113–127.

Gezon, Z. J., E. S. Wyman, J. S. Ascher, D. W. Inouye, and R. E. Irwin. "The Effect of Repeated, Lethal Sampling on Wild Bee Abundance and Diversity." *Methods in Ecology and Evolution* 6 (2015): 1044–1054.

Gianquitto, T. *"Good Observers of Nature": American Women and the Scientific Study of the Natural World*. Athens: University of Georgia Press, 2007.

Gibbons, A. "Spotting Evolution among Us." *Science* 363 (2019): 21–23.

Gobster, P. H. "Alternative Approaches to Urban Natural Areas Restoration: Integrating Social and Ecological Goals." In *Forest Landscape Restoration: Integrating Natural and Social Sciences*, ed. J. Stanturf et al., 155–176. Dordrecht: Springer, 2012.

Gould, E. A., X. de Lamballerie, P. M. Zanotto, and E. C. Holmes. "Origins, Evolution, and Vector/Host Coadaptations within the Genus *Flavivirus*." *Advances in Virus Research* 59 (2003): 277–314.

Greene, A., and N. L. Breisch. "Caverns of Concrete: Urban Karst Is a Challenging Frontier of Urban Entomology." *American Entomologist* 67 (2021): 40–47.

Griffin, D. R., F. A. Webster, and C. R. Michael. "The Echolocation of Flying Insects by Bats." *Animal Behaviour* 8 (1960): 141–154.

Grilo, C., L. Borda-de-Água, P. Beja, E. Goolsby, K. Soanes, A. le Roux, E. Koroleva, F. Z. Ferreira, S. A. Gagné, Y. Wang, and M. González-Suárez. "Conservation Threats from Roadkill in the Global Road Network." *Global Ecology and Biogeography* 30 (2021): 2200–2210.

Grilo, C., E. Koroleva, R. Andrášik, M. Bíl, and M. González-Suárez. "Roadkill Risk and Population Vulnerability in European Birds and Mammals." *Frontiers in Ecology and the Environment* 18 (2020): 323–328.

Grinnell, J. May 5 notes, *Field Notes*, 1323, Section 2: Death Valley, Calif. 1920. Cited in G. Kroll, "An Environmental History of Roadkill: Road Ecology and the Making of the Permeable Highway." *Environmental History* 20 (2015): 4–28.

Habel, J. C., L. Rasche, U. A. Schneider, J. O. Engler, E. Schmid, D. Rödder, S. T. Meyer, N. Trapp, R. Sos del Diego, H. Eggermont, and L. Lens. "Final Countdown for Biodiversity Hotspots." *Conservation Letters* 12 (2019): art. e12668.

Hamilton, G. C., J. J. Ahn, W. Bu, T. C. Leskey, A. L. Nielsen, Y.-L. Park, W. Rabitsch, and K. A. Hoelmer. *Halyomorpha halys* (Stål). In *Invasive Stink Bugs and Related Species (Pentatomoidea): Biology, Higher Systematics, Semiochemistry, and Management*, ed. J. E. McPherson, 243–292. Routledge, 2018.

Hattam, J. "In Istanbul's Ancient Gardens, a Battle for Future Harvests." *Yale Environment 360* (online), February 25, 2016, https://e360.yale.edu.

Hawks, J., E. T. Wang, G. M. Cochran, H. C. Harpending, and R. K. Moyzis. "Recent Acceleration of Human Adaptive Evolution." *Proceedings of the National Academy of Sciences* 104 (2007): 20753–20758.

Heiling, A. M. "Why Do Nocturnal Orb-Web Spiders (Araneidae) Search for Light?" *Behavioral Ecology and Sociobiology* 46 (1999): 43–49.

Helden, A. J., and S. R. Leather. "Biodiversity on Urban Roundabouts—Hemiptera, Management and the Species-Area Relationship. *Basic & Applied Ecology* 5 (2004): 367–377.

Hervé, V., B. Leroy, A. Da Silva Pires, and P. J. Lopez. "Aquatic Urban Ecology at the Scale of a Capital: Community Structure and Interactions in Street Gutters." *ISME Journal* 12 (2018): 253–266.

Hervé, V., and P. J. Lopez. "Analysis of Interdomain Taxonomic Patterns in Urban Street Mats." *Environmental Microbiology* 22 (2020): 1280–1293.

Hiemstra, A. F., B. Gravendeel, and M. Schilthuizen. "Birds Using Artificial Plants as Nesting Material." *Behaviour* 159 (2021): 193–205.

Hiemstra, A. F., C. W. Moeliker, B. Gravendeel, and M. Schilthuizen. "Bird Nests Made from Anti-Bird Spikes." *Deinsea* 21 (2023): 17–25.

Hiemstra, A. F., L. Rambonnet, B. Gravendeel, and M. Schilthuizen. "The Effects of COVID-19 Litter on Animal Life." *Animal Biology* 71 (2021): 215–231.

Hill, J. E., T. L. DeVault, and J. L. Belant. "A Review of Ecological Factors Promoting Road Use by Mammals." *Mammal Review* 51 (2021): 214–227.

Holthuis, L. B. "Notities betreffende Limburgse Crustacea. I. *Atyaephyra desmarestii* (Millet). II. *Niphargus aquilex* Schioedte." *Natuurhistorisch Maandblad* 39 (1956): 125–129.

Horváth, G., M. Blahó, Á. Egri, G. Kriska, I. Seres, and B. Robertson. "Reducing the Maladaptive Attractiveness of Solar Panels to Polarotactic Insects." *Conservation Biology* 24 (2010): 1644–1653.

Horváth, G., P. Malik, G. Kriska, and H. Wildermuth. "Ecological Traps for Dragonflies in a Cemetery: The Attraction of *Sympetrum* Species (Odonata: Libellulidae) by Horizontally Polarizing Black Gravestones." *Freshwater Biology* 52 (2007): 1700–1709.

Horváth, G., and J. Zeil. "Kuwait Oil Lakes as Insect Traps. *Nature* 379 (1996): 303–304.

Howell, B. K., K. M. Winchell, and T. J. Hagey. "Geometric Morphometrics Reveal Shape Differences in the Toes of Urban Lizards." *Integrative Organismal Biology* 4 (2022): art. obac028.

Ibisch, P. L., M. T. Hoffmann, S. Kreft, G. Pe'er, V. Kati, L. Biber-Freudenberger, D. A. DellaSala, M. M. Vale, P. R. Hobson, and N. Selva. "A Global Map of Roadless Areas and Their Conservation Status." *Science* 354 (2016): 1423–1427.

Jagiello, Z., Ł. Dylewski, and M. Szulkin. "The Plastic Homes of Hermit Crabs in the Anthropocene." *Science of the Total Environment*, 2024: 168959.

Jensen, D., L. Keith, and M. Wilbert. *Bright Green Lies: How the Environmental Movement Lost Its Way and What We Can Do about It.* Rhinebeck, NY: Monkfish, 2021.

Johnsen, A. R., B. Styrishave, and J. Aamand. "Quantification of Small-Scale Variation in the Size and Composition of Phenanthrene-Degrader Populations and PAH Contaminants in Traffic-Impacted Topsoil." *FEMS Microbiology Ecology* 88 (2014): 84–93.

Kelly, J. "Inside the Tunnels of Washington's Mole Man, Harrison G. Dyar." *Washington Post*, November 3, 2012.

Kennedy, M. "McEwan's Pebbles Back on Beach." *Guardian*, April 7, 2007.

Kerstes, N. A., T. Breeschoten, V. J. Kalkman, and M. Schilthuizen. "Snail Shell Colour Evolution in Urban Heat Islands Detected via Citizen Science." *Communications Biology* 2 (2019): 264.

Kiew, R., and R. Abdul Rahman. "Plant Diversity Assessment of Karst Limestone: A Case Study of Malaysia's Batu Caves." *Nature Conservation* 44 (2021): 21–49.

Kingsolver, B. *Unsheltered: A Novel.* New York: Harper, 2018.

Klump, B. C., R. E. Major, D. R. Farine, J. M. Martin, and L. M. Aplin. "Is Bin-Opening in Cockatoos Leading to an Innovation Arms Race with Humans?" *Current Biology* 32 (2022): R897–R911.

Klump, B. C., J. M. Martin, S. Wild, J. K. Hörsch, R. E. Major, and L. M. Aplin. "Innovation and Geographic Spread of a Complex Foraging Culture in an Urban Parrot." *Science* 373 (2021): 456–460.

Knutson, R. M. *Flattened Fauna: A Field Guide to Common Animals of Roads, Streets and Highways.* Berkeley, CA: Ten Speed Press, 1987.

Kolenda, K., M. Pawlik, N. Kuśmierek, A. Smolis, and M. Kadej. "Online Media Reveals a Global Problem of Discarded Containers as Deadly Traps for Animals." *Scientific Reports* 11 (2021): 1–10.

Kovtun, O. A. "Подземные сооружения Одессы и Одесской области: Сборник материалов 1-й научно-практической конференции (Одесса 11–12 ноября 2017 г.) [Underground structures of Odessa and Odessa Region: Collection of the materials of the 1st Scientific and Practical Conference (Odessa 11–12 November 2017]." Одесса, 2017.

Kovtun, O. A., R. S. Vargovitsh, M. O. Son, and I. A. Balashov. "Invasive Land Snail *Oxychilus translucidus* (Stylommatophora, Zonitidae) in the Catacombs of Odesa (Ukraine)." *Vestnik zoologii* 51 (2017): 353–354.

Kowarik, I. "Novel Urban Ecosystems, Biodiversity, and Conservation." *Environmental Pollution* 159 (2011): 1974–1983.

Kraemer, M. U. G., M. E. Sinka, K. A. Duda, A. Q. N. Mylne, F. M. Shearer, C. M. Barker, C. G. Moore, et al. "The Global Distribution of the Arbovirus Vectors *Aedes aegypti* and *Ae. albopictus*." *eLife* 4 (2015): e08347.

Kriska, G., Z. Csabai, P. Boda, P. Malik, and G. Horváth. "Why do Red and Dark-Coloured Cars Lure Aquatic Insects? The Attraction of Water Insects to Car Paintwork Explained by Reflection–Polarization Signals." *Proceedings of the Royal Society B* 273 (2006): 1667–1671.

Kriska, G., G. Horváth, and S. Andrikovics. "Why Do Mayflies Lay Their Eggs En Masse on Dry Asphalt Roads? Water-Imitating Polarized Light Reflected from Asphalt Attracts Ephemeroptera." *Journal of Experimental Biology* 201 (1998): 2273–2286.

Kutschera, U., and W. R. Briggs. "From Charles Darwin's Botanical Country-House Studies to Modern Plant Biology." *Plant Biology* 11 (2009): 785–795.

Lang, D. "The Digital Naturalist." *Open Explorer Journal*, April 24, 2015, medium.com/openexplorer-journal.

Laurance, W. "If You Can't Build Well, Then Build Nothing at All." *Nature* 563 (2018): 295–296.

Lawton, G. "Road Kill." *New Scientist* 239, no. 3193 (2018): 36–41.

Lebreton, F., and S. Héas. "La spéléologie urbaine: Une communauté secrète de cataphiles." *Ethnologie française* 37 (2007): 345–352.

Lehner, P. N. "Avian Davian Behavior." *Wilson Bulletin* 100 (1988): 293–294.

Lehtonen, T. K., N. L. Babic, T. Piepponen, O. Valkeeniemi, A. M. Borshagovski, and A. Kaitala. "High Road Mortality during Female-Biased Larval Dispersal in an Iconic Beetle." *Behavioral Ecology and Sociobiology* 75 (2021): 1–10.

Li, F., R. Wang, J. Paulussen, and X. Liu. "Comprehensive Concept Planning of Urban Greening Based on Ecological Principles: A Case Study in Beijing, China." *Landscape and Urban Planning* 72 (2005): 325–336.

Lim, T. W., and T. W. Lim. "Hachioji's Environment." In *History and Regional Area Studies of Hachioji: Tokyo's Western Frontier*, 81–96. New York: Palgrave Macmillan, 2021.

Lima, A., D. D. Lovin, P. V. Hickner, and D. W. Severson. "Evidence for an Overwintering Population of *Aedes aegypti* in Capitol Hill Neighborhood, Washington, DC." *American Journal of Tropical Medicine and Hygiene* 94 (2016): 231–235.

Littleford-Colquhoun, B. L., C. Clemente, M. J. Whiting, D. Ortiz-Barrientos, and C. H. Frère. "Archipelagos of the Anthropocene: Rapid and Extensive Differentiation of Native Terrestrial Vertebrates in a Single Metropolis." *Molecular Ecology* 26 (2017): 2466–2481.

Longino, J. T., and D. B. Booher. "Expect the Unexpected: A New Ant from a Backyard in Utah." *Western North American Naturalist* 79 (2019): 496–499.

Losos, J. B., and C. J. Schneider. "*Anolis* Lizards." *Current Biology* 19 (2009): R316–R318.

Loss, S. R., T. Will, and P. P. Marra. "Estimation of Bird-Vehicle Collision Mortality on U.S. Roads." *Journal of Wildlife Management* 78 (2014): 763–771.

Malik, P., R. Hegedüs, G. Kriska, and G. Horváth. "Imaging Polarimetry of Glass Buildings: Why Do Vertical Glass Surfaces Attract Polarotactic Insects?" *Applied Optics*, 47 (2008): 4361–4374.

Mammola, S., M. Isaia, D. Demonte, P. Triolo, and M. Nervo. "Artificial Lighting Triggers the Presence of Urban Spiders and Their Webs on Historical Buildings." *Landscape and Urban Planning* 180 (2018): 187–194.

Marín-Gómez, O. H., C. Rodríguez Flores, and M. del Coro Arizmendi. "Assessing Ecological Interactions in Urban Areas Using Citizen Science Data: Insights from Hummingbird–Plant Meta-Networks in a Tropical Megacity." *Urban Forestry & Urban Greening* 74 (2022): 127658.

Martin, R. A., L. D. Chick, M. L. Garvin, and S. E. Diamond. "In a Nutshell, a Reciprocal Transplant Experiment Reveals Local Adaptation and Fitness Trade-offs in Response to Urban Evolution in an Acorn-Dwelling Ant." *Evolution* 75 (2021): 876–887.

Martin, R. A., L. D. Chick, A. R. Yilmaz, and S. E. Diamond. "Evolution, Not Transgenerational Plasticity, Explains the Adaptive Divergence of Acorn Ant Thermal Tolerance across an Urban–Rural Temperature Cline." *Evolutionary Applications* 12 (2019): 1678–1687.

Mbugua, S. W., C. H. Wong, and S. Ratnayake. "Effects of Artificial Light on the Larvae of the Firefly *Lamprigera sp.* in an Urban City Park, Peninsular Malaysia." *Journal of Asia-Pacific Entomology* 23 (2020): 82–85.

McFarlane, R. "The Invisible City beneath Paris." *New Yorker*, May 23, 2019.

Meshaka, W. E. Jr. "Anuran Davian Behavior: A Darwinian Dilemma." *Florida Scientist* 59 (1996): 74–75.

Messenger, S. "Trillions of Insects Killed by Cars Every Year, Says Study." *Treehugger*, October 11, 2018, https://www.treehugger.com/trillions-of-insects-killed-by-cars-every-year-says-study-4857578.

Milot, E., and S. C. Stearns. "Selection on Humans in Cities." In *Urban Evolutionary Biology*, ed. M. Szulkin, J. Munshi-South, and A. Charmantier, 268–288. Oxford University Press, 2020.

Moates, G. "Small Mammal Mortality in Discarded Bottles and Drinks Cans: A Norfolk-Based Field Study in a Global Context." *Journal of Litter and Environmental Quality* 2 (2018): 5–13.

Moeliker, K. *De eendenman*. Nieuw Amsterdam, 2009.

Mörzer Bruijns, M. F. "Faunasterfte door verkeer." *De Levende Natuur* 62 (1959): 73–77.

Muñoz, P. T., F. P. Torres, and A. G. Megías. "Effects of Roads on Insects: A Review." *Biodiversity and Conservation* 24 (2015): 659–682.

Nozeman, C., and J. C. Sepp. *Nederlandsche Vogelen; Volgens Hunne Huishouding, Aert, en Eigenschappen Beschreeven*. Amsterdam: Sepp, 1710.

Oliveira, M., J. Ferreira, V. Fernandes, A. Sakuntabhai, and L. Pereira. "Host Ancestry and Dengue Fever: From Mapping of Candidate Genes to Prediction of Worldwide Genetic Risk." *Future Virology* 13 (2018): 647–655.

Oskam, P. I. J., and J. A. Mota. "Design in the Anthropocene: Intentions for the Unintentional." In *Advances in Design and Digital Communication Digicom,* ed. N. Martins and D. Brandão, 269–279. Berlin: Springer, 2020.

Ożgo, M., T. S. Liew, N. B. Webster, and M. Schilthuizen. "Inferring Microevolution from Museum Collections and Resampling: Lessons Learned from *Cepaea*." *PeerJ* 5 (2017): e3938.

Ożgo, M., and M. Schilthuizen. "Evolutionary Change in *Cepaea nemoralis* Shell Colour over 43 Years." *Global Change Biology* 18 (2012): 74–81.

Peeters, N., and T. van Dijk. *Darwins engelen: Vrouwelijke geleerden in de tijd van Charles Darwin*. Amsterdam: Atlas Contact, 2018.

Pieters, T. "Een blinde vlokreeft maakte als 'grottenbeest' furore in een put." *Leidsche Courant*, May 17, 1983, 5.

Pişkin, T. "'This Is a Murder': Construction Equipment Enters İstanbul's Validebağ Grove." *Bianet*, September 21, 2021.

Plaisier, K. "De collectiebeheerders aflevering 8: Bernhard van Vondel." *Straatgras* 8 (1996): 38–40.

Pocock, M. J., and D. M. Evans. "The Success of the Horse-Chestnut Leaf-Miner, *Cameraria ohridella*, in the UK Revealed with Hypothesis-Led Citizen Science." *PloS One* 9 (2014): art. e86226.

Poulson, T. L., and W. B. White. "The Cave Environment." *Science* 165 (1969): 971–981.

Praveenraj, J., T. Thackeray, A. Mohapatra, and A. Pavan-Kumar. "*Rakthamichthys mumba*, a New Species of Hypogean Eel (Teleostei: Synbranchidae) from Mumbai, Maharashtra, India." *Aqua International Journal of Ichthyology* 27 (2021): 93–102.

Prieto-Barajas, C. M., E. Valencia-Cantero, and G. Santoyo. "Microbial Mat Ecosystems: Structure Types, Functional Diversity, and Biotechnological Application." *Electronic Journal of Biotechnology* 31 (2018): 48–56.

Quitmeyer, A. "Digital Naturalism: Interspecies Performative Tool Making for Embodied Science." In *Proceedings of the 2013 ACM conference on Pervasive and Ubiquitous Computing,* 325–330. New York: ACM, 2013.

Quitmeyer, A., C. C. S. Liem, and J. Huber. "Multidisciplinary Column: An Interview with Andrew Quitmeyer." *ACM SIGMM Records* 10, no. 3 (2018).

Quitmeyer, A., H. Perner-Wilson, and B. Fisher. *Hacking the Wild: Making Sense of Nature in the Madagascar Jungle*. Digital Naturalism and others (open source), May 2015, https://archive.org/details/quitmeyer-perner-wilson-fisher-hacking-the-wild-making-sense-of-nature-in-the-madagascar-jungle-1/page/n115/mode/2up.

Romiti, F., E. Pietrangeli, C. Battisti, and G. M. Carpaneto. "Quantifying the Entrapment Effect of Anthropogenic Beach Litter on Sand-Dwelling Beetles According to the EU Marine Strategy Framework Directive." *Journal of Insect Conservation* 25 (2021): 441–452.

Rose, N. H., M. Sylla, A. Badolo, J. Lutomiah, D. Ayala, O. B. Aribodor, N. Ibe, et al. "Climate and Urbanization Drive Mosquito Preference for Humans." *Current Biology* 30 (2020): 3570–3579.

Rosenzweig, M. L. *Species Diversity in Space and Time*. Cambridge: Cambridge University Press, 1995.

Rosenzweig, M. L. *Win-Win Ecology: How the Earth's Species Can Survive in the Midst of Human Enterprise*. Oxford: Oxford University Press, 2003.

Santilli, L., S. A. Castro, J. A. Figueroa, N. Guerrero, C. Ray, M. Romero-Mieres, G. Rojas, and N. Lavandero. "Exotic Species Predominates in the Urban Woody Flora of Central Chile." *Gayana Botánica* 75 (2018): 568–588.

Santos, S. M., F. Carvalho, and A. Mira. "How Long Do the Dead Survive on the Road? Carcass Persistence Probability and Implications for Road-kill Monitoring Surveys." *PLoS ONE* 6 (2011): e25383.

Schermer, M., and L. Hogeweg. "Supporting Citizen Scientists with Automatic Species Identification Using Deep Learning Image Recognition Models." *Biodiversity Information Science and Standards* 2 (2018): https://doi.org/10.3897/biss.2.25268.

Schilthuizen, J. G. "Determineren met boraxparels." *Gea* 12 (1979): 105–107.

Schilthuizen, J. G. "Kristallen tekenen met uw computer." *Gea* 20 (1987): 82–85.

Schilthuizen, J. G. "Ponskaarten als hulpmiddel bij het determineren van mineralen." *Gea* 11 (1978): 40–41.

Schilthuizen, J. G. "Steenzaag met vertikaal draaiend blad." *Gea* (1981): 151–157.

Schilthuizen, J. G. "Zelfbouw van een Geigerteller." *Gea* 12 (1979): 56–59.

Schilthuizen, M. *Darwin Comes to Town: How the Urban Jungle Drives Evolution.* London: Quercus; New York: Picador, 2018.

Schilthuizen, M. "Carpoelen." *Bionieuws,* June 30, 2006.

Schilthuizen, M. "In de ban van het kleine." *Intermediair* 31, no. 43 (1995): 25–27.

Schilthuizen, M. *The Loom of Life: Unravelling Ecosystems.* Berlin: Springer, 2008.

Schilthuizen, M. "Rapid, Habitat-Related Evolution of Land Snail Colour Morphs on Reclaimed Land." *Heredity* 110 (2013): 247–252.

Schilthuizen, M., T. S. Liew, T. H. Liew, P. Berlin, J. P. King, and M. Lakim. "Species Diversity Patterns in Insular Land Snail Communities of Borneo." *Journal of the Geological Society* 170 (2013): 539–545.

Schilthuizen, M., and N. Peeters. "Mary Lua Adelia Treat." In *Darwins engelen: Vrouwelijke geleerden in de tijd van Charles Darwin,* ed. N. Peeters and T. van Dijk, 117–138. Amsterdam: Atlas Contact, 2018.

Schilthuizen, M., L. P. S. Pimenta, Y. Lammers, P. J. Steenbergen, M. Flohil, N. G. Beveridge, P. T. van Duijn, M. M. Meulblok, N. Sosef, R. van de Ven, and R. Werring. "Incorporation of an Invasive Plant into a Native Insect Herbivore Food Web." *PeerJ* 4 (2016): art. e1954.

Schilthuizen, M., E. M. Rutten, and M. Haase. "Small-Scale Genetic Structuring in a Tropical Cave Snail and Admixture with its Above-ground Sister Species." *Biological Journal of the Linnean Society* 105 (2012): 727–740.

Schilthuizen, M., and F. Vonk. *Wie wat bewaart: Twee eeuwen Nederlandse natuurhistorie.* Amsterdam: Het Spectrum, 2020.

Schiner, J. R. "Fauna der Adelsberger-, Luegger-, und Magdalenen Grotte." In *Die Grotten und Höhlen von Adelsberg, Lueg, Planina und Laas,* ed. A. Schmidl, 231–272. Vienna: Braunmüller, 1854.

Schlaepfer, M. A., M. C. Runge, and P. W. Sherman. "Ecological and Evolutionary Traps." *Trends in Ecology and Evolution* 17 (2002): 474–480.

Schlegel, G. "Levensschets van Hermann Schlegel." *Jaarboek Koninklijke Academie van Wetenschappen,* 1884: 1–97.

Segal, S. *Ecological Notes on Wall Vegetation.* The Hague: Junk, 1969.

Seiler, A., and J.-O. Helldin. "Mortality in Wildlife due to Transportation." In *The Ecology of Transportation: Managing Mobility for the Environment,* ed. J. Davenport and J. L. Davenport, 165–189. Springer, 2006.

Semel, B., and P. W. Sherman. "Intraspecific Parasitism and Nest-Site Competition in Wood Ducks." *Animal Behaviour* 61 (2001): 787–803.

Sergio, F., J. Blas, G. Blanco, A. Tanferna, L. López, J. A. Lemus, and F. Hiraldo. "Raptor Nest Decorations Are a Reliable Threat against Conspecifics." *Science* 331 (2011): 327–330.

Sidorov, D. A., and O. A. Kovtun. "*Synurella odessana* sp. n. (Crustacea, Amphipoda, Crangonyctidae): First Report of a Subterranean Amphipod from the Catacombs of Odessa and Its Zoogeographic Importance." *Subterranean Biology* 15 (2015): 11–27.

Sierra, B., P. Triska, P. Soares, G. Garcia, A. B. Perez, E. Aguirre, M. Oliveira, B. Cavadas, B. Regnault, M. Alvarez, and D. Ruiz. "*OSBPL10*, *RXRA* and Lipid Metabolism Confer African-Ancestry Protection against Dengue Haemorrhagic Fever in Admixed Cubans." *PLoS Pathogens* 13 (2017): e1006220.

Simmons, K. E. L. "Bizarre Behaviour and Death of Male House Sparrow." *British Birds* 78 (1985): 243–244.

Smith, H. L. "Protesters Sit Tight as President Erdogan's Trucks Roll In to Pave Over Validebag Grove in Istanbul." *The Times (London)*, October 3, 2021.

Smith, J. C. "How to Be an 'Urban Naturalist' (& Save the Planet)." *Elephant Journal*, March 20, 2019.

Smith, R. P. "The Bizarre Tale of the Tunnels, Trysts and Taxa of a Smithsonian Entomologist." *Smithsonian Magazine*, May 13, 2016.

Stoner, D. "The Toll of the Automobile." *Science* 61 (1925): 56–57.

Suárez-Rodríguez, M., I. López-Rull, and C. Macías Garcia. "Incorporation of Cigarette Butts into Nests Reduces Nest Ectoparasite Load in Urban Birds: New Ingredients for an Old Recipe?" *Biology Letters* 9 (2013): 20120931.

Talbot, M. "The Rogue Experimenters." *New Yorker*, May 18, 2020.

Tasseron, P., H. Zinsmeister, L. Rambonnet, A. F. Hiemstra, D. Siepman, and T. van Emmerik. "Plastic Hotspot Mapping in Urban Water Systems." *Geosciences* 10 (2020): 342.

Tenopir, C., and D. W. King. "The Growth of Journals Publishing." In *The Future of the Journal*, 2nd ed., ed. B. Cope and A. Phillips, 159–178. Chandos, 2014.

Tjørve, E. "How to Resolve the SLOSS Debate: Lessons from Species-Diversity Models." *Journal of Theoretical Biology* 264 (2010): 604–612.

Treat, M. *Home Studies in Nature*. New York: Harper & Brothers, 1885.

Tresch, J. *The Reason for the Darkness of the Night: Edgar Allan Poe and the Forging of American Science*. New York: Farrar, Straus and Giroux, 2021.

Van Achterberg, K., M. Schilthuizen, M. van der Meer, R. Delval, C. Dias, M. Hoynck, H. Köster, R. Maarschall, N. Peeters, P. Venema, and R. Zaremba. "A New Parasitoid Wasp, *Aphaereta vondelparkensis* sp. n. (Braconidae, Alysiinae), from a City Park in the Centre of Amsterdam." *Biodiversity Data Journal* 8 (2020): art. e49017.

Van Heezik, Y., A. Smyth, and R. Mathieu. "Diversity of Native and Exotic Birds across an Urban Gradient in a New Zealand City." *Landscape and Urban Planning* 87 (2008): 223–232.

Van der Wiel, P. *Welke kever is dat?* Zutphen: Thieme, 1954.

Van Vondel, B. "Another Case of Water Beetles Landing on a Red Car Roof." *Latissimus* 10 (1998): 29.

Vervoort, W., C. Smeenk, C. H. J. M. Fransen, and P. K. L. Ng. "Personal Recollections of Lipke Bijdeley Holthuis." *Crustaceana Monographs* 14 (2010): 77–99.

Viles, H. A., and A. A. Gorbushina. "Soiling and Microbial Colonisation on Urban Roadside Limestone: A Three-Year Study in Oxford, England." *Building and Environment* 38 (2003): 1217–1224.

Wang, K., Z. B. Yu, G. Vogel, and J. Che. "Contribution to the Taxonomy of the Genus *Lycodon* H. Boie in Fitzinger, 1827 (Reptilia: Squamata: Colubridae) in China, with Description of Two New Species and Resurrection and Elevation of *Dinodon septentrionale chapaense* Angel, Bourret, 1933." *Zoological Research* 42 (2021): 62–86.

Williams, A. T., and N. Rangel-Buitrago. "The Past, Present, and Future of Plastic Pollution." *Marine Pollution Bulletin* 176 (2022): 113429.

Winchell, K. M., I. Maayan, J. R. Fredette, and L. J. Revell. "Linking Locomotor Performance to Morphological Shifts in Urban Lizards." *Proceedings of the Royal Society B* 285 (2018): 20180229.

Wolters, M., J. Geertsema, E. R. Chang, R. M. Veeneklaas, P. D. Carey, and J. M. Bakker. "Astroturf Seed Traps for Studying Hydrochory." *Functional Ecology* 18 (2004): 141–147.

Xie, T. "Meet the DIY Diggers Who Can't Stop Making 'Hobby Tunnels.'" *Bloomberg News* (online), February 3, 2024.

Yin, S. "What Makes a City Ant? Maybe Just 100 Years of Evolution." *New York Times*, April 3, 2017, https://www.nytimes.com/2017/04/03/science/acorn-ants-evolution-cleveland.html.

INDEX